Explaining Technical Change in a Small Country

The Finnish National Innovation System

ETLA - The Research Institute of the Finnish Economy
Series B84

The Research Institute of the Finnish Economy (ETLA) was founded in 1946 to conduct research in the fields of economics, business and social policy designed to serve financial and economic-policy decision making. At present the members of the association sponsoring ETLA are the central association of industry and employers, major banks and the central association of insurance companies. ETLA also conducts special studies financed from sources outside the association, for which a separate unit has been established. ETLA publishes monographs, reviews and forecasts in several different series.

Address: ETLA, Lönnrotinkatu 4 B, SF-00120 Helsinki, Finland
Telephone 359-0-609900
Telefax 358-0-601 753

Synnöve Vuori
Pentti Vuorinen (Eds.)

Explaining Technical Change in a Small Country

The Finnish National Innovation System

With 30 Figures

Physica-Verlag
in Association with ETLA, Helsinki

Synnöve Vuori
Pentti Vuorinen

ETLA
The Research Institute of the Finnish Economy
Lönnrotinkatu 4 B
SF-00120 Helsinki
Finland

ISBN 978-3-7908-0760-8 ISBN 978-3-642-95913-4 (eBook)
DOI 10.1007/978-3-642-95913-4

88/7130-5 4 3 2 1 0 - Printed on acid-free paper

FOREWORD

The idea to produce this book was based on the observation that technical change in small countries has been rather well described, but not very well explained. In addition, it is based on recognising that technical change is produced by the interaction of a large number of technical, economic, social and institutional factors. It was thus quite natural to take the concept of national innovation systems, as formulated by Bengt-Åke Lundvall and used by several other researchers, as a key concept for the book. We wanted to take Finland as an illustration of more general developments, but also to include international comparisons to allow more general conclusions to be drawn on the basis of the material to be produced. We thus invited several researchers and experts of various fields to give their views on why developments have turned out as they have, and to form a coherent picture of where we are now and where we are heading. Fortunately, we succeeded in getting a well-balanced group of prominent experts to take part in this project. We would like to thank the authors warmly for their contributions.

Most of the papers in this volume were presented at a seminar held on February 2, 1993 in Helsinki. That occasion produced a lively, and, in our view, fruitful discussion about matters that are highly relevant for technical change. The participants were, even though representing various disciplines and lines of thinking, really discussing the same problems and issues, not just presenting their own views. For this we would like to thank the authors and other seminar participants.

We are indebted to several persons for help in producing this book. Pekka Ylä-Anttila was one of the initiators of the project, and an informal group consisting of Markus Koskenlinna, Tarmo Lemola, Raimo Lovio, Erkki Ormala and Pekka Ylä-Anttila, in addition to the editors, discussed the goals and contents that would be relevant for it. Timo Lepistö translated one of the chapters, and Kristina Puranen and John Rogers checked and improved the language of the others. Maarit Säynevirta assisted with data and editing tasks, and the figures were drawn by Arja Selvinen. Tuula Ratapalo and Elina Mikkelä produced the final layout of the book. Thanks are also due to Martti Mäenpää for useful comments at the above-mentioned seminar. Financial support from The Technology Development Centre (TEKES) is gratefully acknowledged.

Helsinki, December 1993

Synnöve Vuori *Pentti Vuorinen*

Contents

1 Outlines of the Finnish Innovation System: The Institutional Setup and Performance

Synnöve Vuori and Pentti Vuorinen
ETLA, The Research Institute of the Finnish Economy, Helsinki, Finland

1.1 Posing the Question

1.1.1 Technical Change and the Size of the Economy

In times of rapid technological change, small countries have specific problems. They are, more than larger ones, affected by the limited availability of resources, by the small size of markets and population, by a limited number of large national carriers of technological know-how, and often also by larger problems of brain drain. They are also more dependent on foreign trade and thus, both more severely hit by protection and more vulnerable to changes in the global competitive scene. Small economies have less opportunities to monitor technical development on a wide front and develop new basic technologies. With growing internationalization and recent structural changes in the world economy, the 'small country squeeze' has even worsened and narrowed the domain in which the smaller countries can operate without major problems (see e.g. van Tulder 1989).

In new technology, there is a tendency towards an increased complexity over all industries which raises entry barriers and makes it more difficult for small economies to compete in the global market. For example, Walsh (1988) argues that the technology of today's car production is approaching the complexity characteristic of yesterday's aircraft industry.

One outcome of this development is that for the effective exploitation of technologies in industry, a much broader spectrum of R&D capability is required. This seems to be true even in sectors with medium- and low-tech products. Consequently, much greater expenditure on human and financial resources is needed.

Hence small countries must either spread their resources more thinly over the available areas, or otherwise select certain areas as priorities for R&D investment. For example, international comparisons of patenting activities reveal that small countries are more narrowly specialised in particular technological fields than large ones (Soete 1988).

However, in spite of the obvious disadvantages of smallness, many small economies have shown remarkable success. This is confirmed, e.g., by the rise of the

Nordic countries from a very modest beginning in the early 20th century among the richest countries in terms of per capita domestic product.

Many authors (e.g.. Tisdell 1982) claim that small economies - e.g. the Nordic countries in comparison with the USA - have reacted promptly to changed world economic conditions and have remained flexible in their science and technology policies. It can thus be argued that it is not the size, but rather adaptability, flexibility and the preparedness to develop appropriate productive structures that are the crucial factors (e.g. Walsh 1988). Consequently, the combined institutional characteristics of a national innovation and production system may have more power to explain technological change than the size or any other directly quantitative economic indicators.

1.1.2 National System of Innovation and Production

According to Edquist and Lundvall (1989), the national system of innovation consists of the institutions and economic structures affecting the rate and direction of innovative activities in the economy. There is no precise model for the system: the boundaries of a national system have to be determined by a concrete analysis combining theoretical and historical issues. The components, internal structures and dynamics of a national innovation and production system change historically.

The weight of the 'national' component also varies. Technological advance has, to a growing extent, become international and global, and the international institutions often play a major role in national development. However, e.g.. Lundvall (1992) emphasizes that nation states have worked as 'engines of growth' by creating the institutional framework and state policies that have supported the rapid economic transformation. Many small nations - Finland especially - still constitute culturally homogeneous systems and offer coherent institutional contexts for socio-economic development.

The elements of a national innovation and production system include e.g. (see Lundvall 1992):

- the internal organization of firms
- inter-firm relationships and economic structures
- role of the public sector
- institutional set up of the financial sector
- R&D intensity and R&D organization
- the national education and training system

Johnson, when discussing technological development as an institutional learning process, formalized the main components of a national innovation and production system as in figure 1.1 (Johnson 1988). In his approach searching and learning are the crucial processes of technological change. The national system creates the context, directing, supporting or restraining these processes.

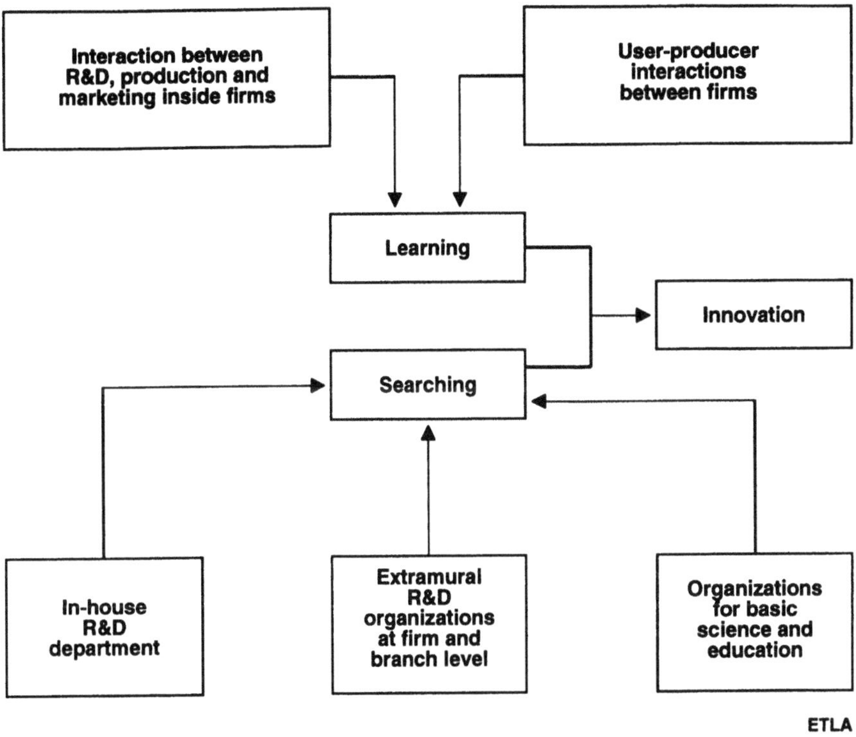

Fig. 1.1. The main components of a national innovation and production system. Source: Johnson 1988.

1.1.3 The Aim of the Book: Explaining Finnish Technological Development

According to many indicators, technological development of the Finnish economy has been rapid (see section 1.3). However, this development has been only vaguely explained. The topic has been the subject of only a few economic studies, and scholars of other scientific disciplines have been even less interested in technological issues.

There are institutional and internal reasons for this. Whereas conventional economic science seems to have relatively poor instruments to grasp the essential dynamics of technological change, more unorthodox approaches are still scarce in Finland. Within history, philosophy and policy studies, technology has neither been a popular subject of study, partly because of the lack of research traditions and a few theoretical explanations and models available in national discussion. Nor is there any research institute specialized in technology studies. Only a few fields - such as industrial sociology - have acquired more comprehensive experience in technology issues. Hence, in terms of technology studies, Finland lags far behind most industrialized countries, although in terms of actual technological development, she has fared well (see e.g. Lemola et al. 1990).

In the present situation, with tightening European integration, rapidly widening international markets and intensifying technological competition, the understanding of the dynamics of technological change has become urgent. It is even more important in the light of the severe conditions of the Finnish economy. A new growth path for the Finnish economy urgently calls for a sound technological basis. In creating this basis, policy measures are crucial. They can only be adequately designed, if the dynamics within the national innovation and production system are known.

The aim of this book is to review and analyze the characteristics and performance of the Finnish innovation system in the last couple of decades, and to assess the development paths possible in the near future. This first article presents some background information on the features of the Finnish innovation and production system (section 1.2) and an analysis of the technological performance and productivity developments of Finnish industry in the last 20 years (section 1.3). The other articles have been written by experts representing various research disciplines. While this first article mainly describes technical change in Finland so far, the experts will explain these developments in the following articles from their own point of view.

1.2 The Institutional Set-Up for Technical Change in Finland

1.2.1 Production Structure and Technological Development

The Background to Development: Channels of Early Technology Transfer. A major share of technology is transferred from abroad and not developed domestically. This is especially true for small industrial latecomers like Finland. Myllyntaus has studied the Finnish model of technology transfer from a historical point of view. He separates eight basic types of technology transfer (table 1.1). During the industrialisation period, the imports of foreign machinery were the most important transfer channel, accompanied by nationals' journeys abroad and natural diffusion. Later on, the role of recruiting foreign experts has declined, while the acquisition of foreign licences and patents became important, especially after World War II (Myllyntaus 1992).

The modest role of direct foreign investment is a special feature of the Finnish model. It seems to be related to the strong tendency of economic nationalism that has also restricted the utilisation of foreign turn key deliveries and joint ventures. Myllyntaus emphasizes that even the postwar development rests on some of the nationalist principles typical of the late 19th century (Myllyntaus 1992). He concludes:

Another peculiarity of the model has been the strong emphasis on promote nationals' role in the transfer and practical application of foreign technology. At the beginning, this meant choosing a long and slow road to industrial and technological modernisation. A basic reason for this was to keep technology transfer under the recipient's control. This could also be called a cornerstone of the Finnish model of technology transfer. (Myllyntaus 1992)

Table 1.1. The significance of different technology transfer channels in Finland from 1809 on

Channels for technology transfer	1809-1917	1918-1944	1945-1992
Direct foreign investments	1[1]	1	1
Importing foreign machinery	3	3	3
Acquiring turn-key plants	0	0	1
Acquiring foreign licenses and patents	1	2	3
Setting up joint ventures with foreigners/ foreign companies	1	0	1
Recruiting skills from abroad	3	2	1
Encouraging nationals' journeys and studying abroad	3	2	3
Natural diffusion through trade, publications etc.	2	3	3

[1] Scale 0 - 3: from unimportant to very important
Source: Myllyntaus 1992

Economic Structure and Technological Development: Development Blocks. Technology is often transferred from abroad; however, it is not planted in virgin soil but in the already existing institutions and production structures. Lundvall emphasizes the role of historical developments even in this respect:

> In most small countries, the present pattern of specialization in exports can be traced far back in their economic history - agro-industrial products in Denmark and Holland, products related to timber and metalworking in Sweden and Finland, products relating to shipping in Norway, etc. (Lundvall 1986)

In the core of this specialization pattern are industrial complexes or 'development blocks' that can be defined as broad sets of interconnected users and producers, developing in close interaction with each other and often supported by private or public organizations producing knowledge. Such a development block typically comprises various domestic sectors coupled with strong linkages, and it is often constituted around a strategic export sector (Edquist & Lundvall 1989). In a national production system development blocks construct the framework for interactive learning and the specific national combination of such blocks is vital in regulating the speed and direction of technological change.

In Finland, the forest sector forms the focal development block. The forest sector has dominated the Finnish industry, and has been able to create dynamic growth with a constantly widening scale of exports. The sector has had an impact on Finnish society extending over the economic dimension. For example, the significance of the forest industries in the formation of communities in Finland has been obvious (see e.g. Raumolin 1990).

Even if the metal and engineering industries have grown much faster since the early 1970s, they have not built up such a consistent complex (figure 1.2). The metal and engineering industries are rather based on producing capital goods for other sectors, being less coherent and having more diversified connections with other complexes. However, the role of the metal and engineering industries has been steadily growing, and, in terms of R&D, it is by far the most important sector (see section 1.3).

The important status of the forest industries is illustrated by their high global market shares. E.g., in many paper product groups Finnish corporations are the leading exporters. The overall market share of Finnish companies in world paper and board exports has remained between 14 and 15% since the 1960s. However, the product structure has changed towards products with a higher value added; e.g., the market share of writing papers grew from 21.5% in 1960 to 30.5% in 1989. Simultaneously, the market share of world pulp exports - largely due to an increasing vertical integration - dropped from 16.6.% in 1960 to 6.5% in 1989 (Lilja et al. 1992).

The success has often been explained by the internal dynamics of the sector, which has been able to mobilize diverse agents with even opposing interests into a

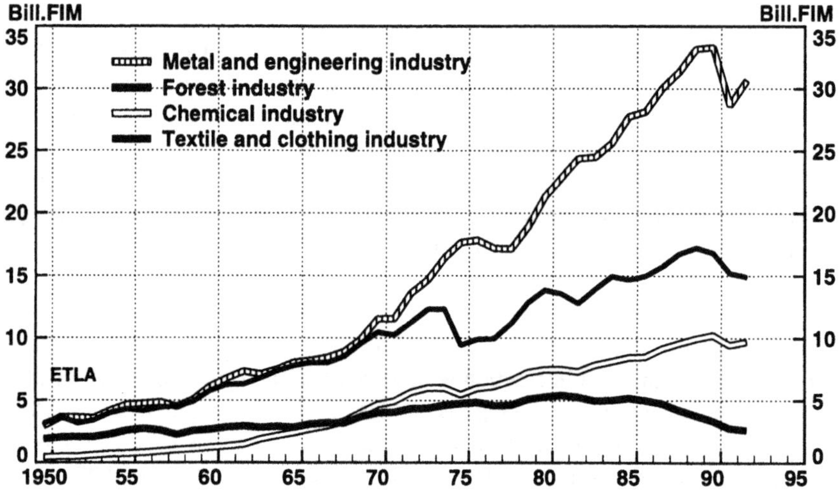

Fig. 1.2. Manufacturing output by industries (value added at 1985 prices). Source: ETLA.

huge collective project. The development block, or a business system, of the forest industries includes, in addition to forestry, the mechanical and chemical forest industries also several supplier and customer industries, which together create a strong entity with intensive user-producer interactions.

Lilja et al. explain the success with a multitude of largely institutional factors within the system. These include the specific corporate specialization pattern, close relations with the banking sector, general technological modernism, consortative business operations, specific labour relations within the sector and distinct linkages to the state (Lilja et al. 1992).

This combination of economic, social and political factors within the sector has also led to rapid technological advance. The mechanisms promoting technology diffusion

> ..include the closeness of the mills to each other, networks derived from the educational institutions, a joint R&D research institute, professional societies, consulting firms, suppliers of machines and systems and career paths from one corporation to another. Companies which have failed in their modernization efforts have found a safety network in the industry because they have been acquired by firms with stronger resources (Lilja et al. 1992).

However, a recent critique points out that the forest sector has not invested much in product innovations. The Finnish forest industries have largely relied on incremental development within the conventional product variety, and they are facing problems with more radical product innovations entering the world market. This includes the revision of attitudes and regulations concerning the environment, e.g., on the share of recycled mass in paper products. This fact is also shown in the low figures of R&D expenditure of the industries: in 1989 the share of R&D expenditures in value added was only 2.2% in pulp, paper and paper products, as against, e.g., 24% in drugs (see section 1.3).

Supplier industries have been technologically more active, and in many types of paper machinery Finnish companies are among world leaders, in respect of both market shares and technical standards. The development only started after the Second World War: the first paper machine was exported (by Wärtsilä/Kone ja Silta) in 1948 to Czechoslovakia, exports took really off during the 1950s, and in the late 1980s, Finland was the world leading exporter of paper machinery (Hjerppe 1990). This is largely an outcome of the integration and intense user-producer co-operation within the sector. The forest industries have bred techno-economic advancement to supplier industries.

It is not confined to the machinery industry only. The linkages go further: also the growth of the electronics industry has been firmly interwoven with the forest sector. Although the birth of the sector is less related to the forest sector, the development of technologically more advanced products is. For example, the automation industry is largely a result of interaction between the manufacturers of electronics and paper machinery. Company histories tell the same story: many electronics firms are spin-offs from forestry/paper machine companies, often remaining within the same group of companies.

The growth of the electronics industry in the 1980s was quite phenomenal, in terms of company internationalization, market developments and technological advance. As a company, Nokia was often taken as an example of the national structural change, with the focus of business moving from the forest industries towards the electronics and telecommunication industries. As a research institute, the company also became the by far largest in Finland during the decade, with the bulk of its activities being directed to developing the electronics (see e.g. Lovio 1989) industry.

However, by the end of the decade the growth had largely reached its peak, and at present the most advanced units have either been sold to multinationals - e.g. computer manufacturing from Nokia to ICL and, as the last step, Alhström automation to Honeywell - or been closed down (Micronas integrated circuit manufacturing). The reasons for this remain open: was the basis for a radical structural change not firm enough, was there not enough technological capability, or were the basic reasons financial? However, in a national perspective, the role of the forest industries seems to be enhanced again.

In addition to the forest industry block, the production chain around the mining industry can be mentioned as a similar type of development block. Like the forest industries the mining industry has bred technical and economic development through intensive linkages to the machinery and electronics industries. However, in this case the international structural change of the complex has proceeded further, while mining activities are increasingly decreasing in Finland. The focus of activities has moved to international operations and the planning and manufacture of mining equipment.

Diffusion of Process Technologies. In the core of the Finnish production system, the forest industries, production technology is among the most modern in the world, and the newest technologies have diffused quite evenly throughout the respective sectors. The situation is quite comparable in some other process industrial sectors, e.g., petroleum refining.

In other sectors of the manufacturing industries the conditions are somewhat different. In fact, with the occasional exception of the textiles and clothing industries the Finnish production structure has nowhere outside the forest industries comprised any real mass production. Product series and production units have, by international standards, been very small.

This is reflected in the utilization of production technologies. Table 1.2 reveals that the diffusion rate of robots - technology typical of producing long product series - has been very low. Even the diffusion of NC machine tools has been modest, but in flexible manufacturing systems the Finnish figures are internationally rather comparable. This observation is often explained by the especially flexible use of the Finnish FM systems; thus they have even been employable in producing batches of very few units only. However, the peak is quite thin, and the most modern technologies are all employed by a few top companies (Ollus et al. 1990; see also Ray 1992).

Table 1.2. The number of flexible manufacturing systems, numerically controlled machine tools and robots in relation to 100 000 industrial workers in some countries in the late 1980s

Country	FMS	NCMTs	Robots
Sweden	3.7	1004	312
Finland	1.9	565	88
United Kingdom	1.7	991	84
France	1.4	1118	145
Japan	1.2	1461	1113
Tsechoslovakia	0.9	...	253
F.R. of Germany	0.9	904	244
USA	0.6	946	153

Source: Ollus et al. 1990

Product Innovations. The technological level of products in a country is often described by patent statistics and statistics on trade in high-tech products. International comparisons of patenting activities reveal that small countries are more narrowly specialised in particular technological fields than large countries (Soete 1988).

This is quite true for Finland as well: the majority of patents are related to the core sectors of industry. For example, the majority of Finnish companies taking patents in the USA in the 1980s were engaged in the machinery, electronics and chemical sectors. However, about one third of all patents were related to the process or machinery of the forest industries. The same statistics show that Finnish patenting in the USA has slowed down since the late 1980s. This was largely due to the decrease in the patenting activities of old companies and old technological fields which has not been compensated by new fields of technology or new innovative firms. (Valkonen & Lovio 1991). Finnish exports of high-tech products are relatively small but growing steadily. The Finnish trade in high-tech and patenting issues will also be discussed in Section 1.3.

1.2.2 Infrastructure and Institutional Support

Skills, Education and Training. Abilities to innovate, implement, improve and use new technologies in existing economic organizations are the basic conditions for all technological advance. Hence education, training and other forms of skills development are crucial issues. Here skills should be viewed from a broad angle; it is neither technical skills alone, the skills acquired formally through education nor individual skills that are important. Technological progress is also implanted in many basic skills, which are often acquired through informal methods, like learning by doing and communicating; furthermore, collective skills embodied in the

Table 1.3. The highest level of education attained in the Nordic Countries as of 1988, by age group, per cent

	Comprehensive education only	Secondary, second stage	Third level
Denmark			
total, 30-64	45.2	36.6	18.3
30-39	36.7	38.0	23.3
60-64	63.2	26.4	10.4
Finland			
total, 30-64	48.5	34.8	16.7
30-39	29.8	49.9	20.3
60-64	77.6	13.7	8.7
Norway			
total, 30-64	46.5	35.3	18.2
30-39	31.4	44.5	24.1
60-64	67.7	23.3	9.0
Sweden			
total, 30-64	43.7	37.1	19.3
30-39	33.4	42.6	24.0
60-64	62.7	27.5	9.8

Source: Statistical reports of the Nordic countries No. 56

organization and cooperative patterns within a working team are often more important than high skills of individual team members.

In respect of creating conditions for technical change through formal education Finland is doing quite well. The Finnish educational system is usually praised for being both comprehensive and proficient. Statistically, everybody leaving the compulsory comprehensive school at the age of 15/16, is guaranteed an opportunity to continue education within the system for secondary education. In practice, approximately 5% of primary-school leavers were left outside the system after compulsory education in the 1980s. However, even some of them continued their education later through apprenticeship or other forms of adult education.

If we look at the whole population, the educational situation is not so good. A Nordic comparison (table 1.3) reveals a drastic difference between age groups. In the younger age groups (30-39), less than 30% - the lowest figure in the Nordic countries - had only compulsory education, while in the oldest age group (60-64) the share was 77.6%. This was clearly the highest Nordic figure.

In the younger age group of 30 to 39, the Finnish figures were especially appreciable in secondary second stage education, which was the highest level attained by almost 50% of the age group. Historically, the figures show that a vast majority of the working population did not have any formal vocational education during the first two or three decades after the Second World War. Formally acquired vocational skills have only become common since the late 1970s (see also Asplund 1991).

It could be argued that the basic period of industrialisation and technological catching-up was rather based on skills acquired through practice and general social factors like economic organization, work values and attitudes than on formal education. Informal training and adult education not shown in the statistics might have played some role. However, vocational adult education in industrial occupations has been very modest up to the 1970s (e.g. Komiteanmietintö 1988:28). The role of informal education and learning by doing is more difficult to prove. One explanation could be searched from the background of industrial workers in the 1950s and 1960s: a majority of them came from an agrarian environment. Was the industrialization and basic technological advance actually based on agrarian skills - and values - converted into a new industrial environment?

Conversely, the 'informatisation', or 'postindustrial' period seems to be much more based on formally acquired skills and vocational education. Or, is it only that Finland has also 'caught up' other industrial countries in terms of education as well?

Later developments in the system of secondary education include increased co-operation between, on one hand, general and vocational secondary education, and, on the other hand, between vocational education institutions and companies. The prevailing system has been widely criticized for being too rigid and formal. For example, the very theoretical emphasis of vocational education from the beginning has often curbed the motivation of practically oriented pupils and led to interruptions of studies (see Ekola et al. 1991).

Generally the situation is very similar in higher education. In the late 1980s, Finland was still lagging behind in the number of third-level degree holders as compared with other Nordic countries. However, if we consider the technologically most important group, science and engineering graduates alone, the picture is entirely different (table 1.4). In 1988, the number of graduates was even, in absolute terms, higher than in any other Nordic country. Also the growth of science - and especially engineering - graduates has accelerated remarkably in the 1980s.

The status of first post-secondary level education can also to be raised by transforming some of the best technical and business education institutes into vocational colleges and combining the resources of previously independent institutes.

Table 1.4. Science and engineering graduates (first post secondary level) per 100.000 persons of working age in the Nordic Countries in 1988

	Graduates/100.000	Number of graduates
Denmark	90	2771
Finland	142	4302
Norway	126	3020
Sweden	84	4086

Source: Statistical reports of the Nordic countries No. 56

Other Infrastructure Issues. One peculiar feature of the Finnish education system is that further education and training abroad have been rare up to the 1980s. This is quite true even for academic studies, but especially for lower grades. Only since the late 1980s policies have put more emphasis on acquiring experience from foreign studies. However, the basic principle that everyone should be guaranteed an opportunity for further studies in the home country has not changed.

In addition to education and training, communication technologies for logistical networks, the financial system and the issues related to social infrastructure institutions are important for technological development. On the other hand, the building-up of infrastructure has, in itself, created technological advance.

From the point of view of initial industrialization and the building-up of the basis for technological advance, networks of physical communication were crucial. Basic networks of roads, railroads and canals - especially important for transporting raw materials for the forest industries - were constructed quite early in the late 19th century.

The creation of telecommunication networks also started relatively early - Finland stands internationally very high, e.g., in the comparison of telephones per capita - and the modernization of the network has taken place quite rapidly. The networks for portable NMT phones are comprehensive, which is related to the fact that Nokia has succeeded globally very well in the manufacture of portable phones. There may be a specific connection between the large area of Finland, the high need for telecommunication, the early building-up of networks and the technical progress of the sector in Finland.

Other basic infrastructure issues include the building-up of the energy supply system, (basically water) power plants and electricity transfer networks. This was, as such, a demanding technological task, and - as Myllyntaus (1991) proves - an important arena for transferring foreign technology in the country and for learning by doing. The same holds largely true for the construction of industrial and residential buildings, although on a lower and less demanding technological level. This was especially intense in the decades of rapid industrialization and internal migration from the agrarian periphery to the south, towns and other more central regions.

Finance is another basic condition for technological development. Typically the share of equity capital has been very low in Finnish industry, and the role of banks has been crucial. The ownership structures of Finnish industrial complexes are very much concentrated around the main banks. The crisis of the early 1990s has revealed the vulnerability of the financial system and ownership structures quite severely.

In respect of technical change, finance has been a problem, in spite of the rapidly expanding public support (see section 1.2.3). Scarce finance hits most severely small high-tech firms developing new products. The more advanced and the more based on basic technology the innovation concerned is, the longer will the development phase before the real market stage be. The period of no profits and high-costs may take years.

Various venture capital and development companies became quite important sources in financing small high-tech firms in the 1980s. For example, their electronics companies employed ca. 4.000 people in the late 1980s, a 10% share of the sector's total employment (Lovio 1990). The recession of the early 1990s has, however, severely hit the activities of venture capital and development companies.

Even large corporations face problems in financing further technological developments. The sales of important divisions to multinational companies is one indicator of this. Does this mean that it is finally impossible for companies in a small country to compete on the edge of high-tech because of the rising costs of technological development? Moreover, is this a sign of an inevitable internationalization of production? In Finland, where foreign industrial capital has been rare, this will mean a drastic change in the development model.

1.2.3 Firm Behaviour: Technology Management, Linkages and Labour Relations

Technology Strategies. Technical advance is not a question of the industrial structure or formal skills, but an outcome of the management strategies and behavioural decisions of existing companies. Technology strategies are only one dimension in companies' overall strategies - technology is just one means of competing, surviving and developing. However, the role of technology seems to be increasing, and the technological aspect is included to a growing share of business issues.

Conventionally, it is typical to divide companies in respect of technology strategy - especially product technology - into defensive or offensive ones. Offensive companies try to stay on the edge of technical advance, ahead of their competitors and conquer new markets through product innovations. Defensive companies carry out only the necessary changes in their products and compete more with production costs. Accordingly, a company with a defensive product innovation strategy may be highly up to date in renewing process technologies.

In his study of product innovation strategies of Finnish firms, Lindell (1989) has divided companies according to whether the main aim of strategy is the existing business area or if the company is aiming at new areas. On this basis, he has constructed four prototypes of companies: the analyzer, the defensive defender, the offensive defender and the prospector.

The analyzer has a combination of old and new areas as its goal being relatively cautious and analytic in making decisions. The defensive defender stays in existing business areas and bases its operations on reacting to the actions of competitors; the strategy often emphasizes drastic short-term reactions. The offensive defender lays great emphasis on both old and new business areas, and bases its strategy on planned long term goals. The prospector is aiming with determination at new business areas, representing usually a new firm in the growth phase (Lindell 1989). A prospector type of strategy is typical of new high-tech firms and relatively new in the Finnish context.

Perhaps a strategy resembling the defensive defender has been the most common among Finnish corporations. To a large extent, the Finnish technological catching up process has also been based on this type of approach. Product innovations are more often introduced as a reaction to competitor's measures than as an active pursuit. However, the introduction of an innovation is often combined with the advancement of process technology and incremental developments in the product.

Lindell's case studies point out that the strategy has often produced poor results. This may be the case even more generally in the present context of rapidly changing technologies and a more uncertain economic environment - where the technology gap to be catched up is small.

The changing environment and needs for more comprehensive technology strategies are among the issues that have led to new organizational solutions within and between companies. Technological progress is, to a growing extent, an issue of organizational cooperation and user - producer interaction. Even for large firms the monitoring of all technical novelties in their business area often demands too much resources. Mastering all relevant technologies is a more or less impossible task. For small firms external resources are even more essential.

Firm Linkages: Networks of Small and Large Firms, Research Institutes and Universities. More than large corporations, small high-tech firms are often regarded as the main carriers of the new technological paradigm. This is not necessarily true: for example, most of high-tech production and the majority of private sector R&D work takes place within large corporations.

Table 1.5. R&D expenditures in industry, by the size of firm, in Denmark, Finland, Norway and Sweden in 1989, %

	Denmark	Finland	Norway	Sweden
5 - 49	10	3	11	1
50 - 99	5	2	71	2
100 - 199	9	2	8	2
200 - 499	15	5	12	5
over 500	61	88	62	90
	100	100	100	100

Source: Nordisk Industrifond 1991

In Nordic comparison, the share of R&D expenditures of firms employing less than 50 persons is remarkably low in Finland and Sweden (Table 1.5). The differences in size structures are actually reverse: 82.0% of industrial firms in Sweden employ less than 50 persons, 82.6% in Norway, 84.0% in Finland and 76.8% in Denmark (Nordisk Industrifond 1991). The figures reveal that the share of high-tech firms of all small firms is very small and that the R&D performed by them is rather insignificant in comparison with the R&D by large firms.

However, small firms are an essential element within the technological dynamics. For example, the interaction between small and large firms in the Finnish electronics sector is well described by Lovio (1989): Small firms have often been behind important innovations. But they do not have sufficient resources to develop the new products up to the market stage. Consequently, firms - or the majority of their shares - are sold to larger companies. From then on, innovations are developed and brought to the market by the large company.

The circular movement, however, continues. New small technology firms are again born as spin-offs from the large corporation, or they are intentionally created by the corporate management by divisionalizing activities. Similar developments are typical in more central and larger industrial countries.

The tight linkages between small high-tech firms and large corporations are mentioned also in a study of firms in the Otaniemi Science Park. 54% of the interviewed (n = 54) small firms had at least weekly contacts with large Finnish corporations. This was the highest figure of all contacts, while the Helsinki University of Technology came second with 33% of the firms having at least weekly contacts there (Kauranen et al. 1991).

These findings support the notion of close and frequent linkages between single firms and between firms and research organizations as a significant dimension of national production and innovation systems. The importance of close linkages to, on one hand, other firms, both as suppliers of services and other supportive functions, buyers and sources of information, and, on the other hand to research organizations as sources of information is emphasized in many studies of small high-tech firms (see e.g. Autio et. al. 1989, Vuorinen et al. 1989). For example, most small high-tech entrepreneurs also have their background either in universities or in older existing companies (ibid.).

The close linkages within technological production have assumed their geographical form in 'new industrial spaces' (see e.g. Scott 1988), well typified by Silicon Valley and other Science/Technology Parks. In Finland, the location of high-tech - or research intensive - production generally follows the existing location patterns of large companies, and concentrates quite clearly within the Turku - Tampere - Helsinki triangle, especially in the south between Helsinki and Turku. These are the economically most active regions in the country, with an abundant supply of services and information linkages.

Nearest to the concept of 'new industrial spaces' comes the Oulu region, 600 km north of Helsinki, with an evident cluster of high-tech production concentrated around the Oulu University and the Oulu Technology Park. The technology park was established in 1982 as the first of its kind in the Nordic countries. All the characteristics of the more famous science park environments also exist here: a mixture of research units of large corporations, small high-tech firms, various applied and basic services, university institutions and a heavy concentration of knowledge/research intensive sectors. The Park also has clearly visible regional effects, with spin-off firms in Oulu and the neighbouring municipalities and in locations further off (Vuorinen et al. 1989).

Similar technology parks have been established in the mid-1980s in six other regions. All parks have been set up in a location near a university, and technology transfer from the university/research institute environment to firms is a central aim of the activities. The tenants in all technology parks consist of small start-up high-tech firms, research units of large corporations, university institutions and various service providers. The role, if not magnitude, of high corporate participation is important. For example, Nokia has already established a unit in every park in an early phase (Vuorinen et al. 1989).

However, it is questionable whether the Finnish 'new industrial spaces' of high technology can ever reach the point of 'the critical mass' needed to create a sustainable growth. The recession has worsened the conditions, and technology parks have incurred their lot of financial problems.

Internationalization. Increasing internationalization is the other changing regional dimension of industrial activities. Historically, technology diffusion from abroad has been important for technical advance in Finland, as mentioned above. However, the role of foreign capital in developing industry has been very modest.

The share of foreign companies has been highest - 40% of employees in 1960 - in the electronics industry. However, by international standards, they have only played a minor, even though a positive role, in the creation and growth of the sector. Their earlier significance lay in setting up companies in Finland, which helped to diversify the industrial structure and served as a channel for technology diffusion. At later stages, their other major contribution was to offer skills, channels and resources for exporting Finnish know-how (Lovio 1992).

Vice versa, the situation has been quite similar. The international operations of Finnish industrial firms have been fairly modest even in the 1970s. Only a few companies had establishments abroad, and a majority of them were sales organizations or representatives. Compared with, e.g. Sweden, productive activities abroad were rare.

The situation has changed radically from both perspectives during and since the 1980s. The internationalization of Finnish companies started in the 1970s and accelerated swiftly:

> Rapid growth is reflected, for example, in the fact that in 1990 the total turnover of Finnish-owned manufacturing companies abroad already amounted to almost FIM 100 billion, or nearly triple the amount five years earlier. 45% of turnover came from EC countries, 33% from the EFTA countries and 19% from North America. In five years, the number of jobs in the companies abroad more than doubled, i.e. from 63 000 to 141 000 employees in 1990. (Kuitunen 1991)

Simultaneously, the main target area of foreign direct investment has shifted from Sweden to the EC, and partly also to the USA. Also the motives for foreign direct investment have diversified. In addition to the traditional motives - promoting marketing, moving manufacturing in the market area, acquiring raw materials or, in some cases, moving production to regions of cheaper production costs - the intention to create linkages to sources of information and to possible users of developed technologies appears often.

The fact that foreign activities are connected with firm-specific know-how is obvious, since companies have expanded abroad primarily horizontally. They have strengthened their main industrial branches through foreign activities. Their objective has been to ensure similar competitive conditions as their competitors, to strengthen their market positions, obtain additional resources and enhance their efficiency by means of labour division between their domestic and foreign production units. (Kuitunen 1991)

Also the share of foreign capital in Finland has grown, although more slowly.

At the end of 1990, the stock of foreign-owned investment in Finland amounted to about FIM 18 billion. The most important investor country, Sweden, accounted for some 50% of the stock of outstanding investment. (Kuitunen 1991)

The most significant flow of foreign investment has taken place through acquisition of Finnish firms. From the technological point of view, the most important foreign operations so far have been the sales of Masa Yards to the Norwegian Kvaerner, Strömberg to the ABB and Nokia Data to the ICL. For the moment, the linking of these firms seems to have positive effects on them. The shipyard is prospering more than for years, Strömberg has expanded its Finnish R&D activities, and the ICL is increasing computer production volumes in the country. Within the ABB, Strömberg has gained a status, global linkages and support for technology development that would have been quite unlikely had the firm stayed under Finnish ownership (see Piironen 1992).

In a global perspective, one of the most important issues of intenationalization is the growth of international R&D cooperation between firms. The concept of 'strategic alliances' has become a slogan: even large firms are cooperating on an international scale to solve the basic problems of technological development together. One of the most recent endeavours is a high-level development programme of manufacturing automation technologies, The Intelligent Manufacturing Systems programme, in which firms and research organizations from Europe, Japan and the USA participate. The programme has a time span of more than a decade and very extensive targets.

The research cooperation activities of Finnish firms have also grown rapidly. At least four forms of cooperation seem to be important. First, firms have created working linkages to international user-producer networks through their 'natural' linkages and business connections. Second, firms have established units monitoring research and technology in environments rich in new innovations. Third, originally Finnish firms have established themselves through foreign acquisitions inside the existing development networks. Fourth, Finnish companies are, to a growing extent, participating in the technology development programmes initiated by the EC and other international organizations.

Labour Relations. Finally, human resources are the most crucial factor in technical advance. Work organization and labour relations are thus also important factors in technical change. It is often argued that the way of using manpower has changed as a result of the introduction of flexible production techniques, a need to rapidly change the product varieties, the competitive need to produce more quality and the need to exploit all innovative resources, even the resources of the shop-floor work force. The argument is that a strict division of labour along the taylorist principles is disappearing, and a new organizational model based on team work, good training and broad work tasks is gaining more ground.

There are plenty of examples of radical changes in work organizations. More autonomous working is often used in technologically advanced companies, and company executives stress the importance of well trained workers. Even decision making is often delegated to lower hierarchical levels, and factories themselves are often regarded rather as networks of relatively independent workshops than rigid hierarchies (see e.g. Ollus et al. 1990).

However, many studies (see e.g. Alasoini 1990) point out that this is far from the current reality. Most of the companies using up-to-date technologies and production philosophies in their general operations and strategies are organized on the old principles. There are also workers not so eager to change the working methods they are used to.

Even more generally, the recession has tarnished the image of 'workers as the most important resource' of an enterprise. Arguments of flexibility, the development of work organizations and working conditions are easily forgotten under the pressure of more urgent problems. Furthermore, for many firms, it is appealing to change the work force at times when the supply of labour is abundant.

On the other hand, the uncertainty of employment and the lack of improvement in working conditions has also deteriorated the other part of the new work models of organization and company spirits. Workers may not identify themselves as easily with the 'firm culture' as was often thought in the 1980s (see e.g. Ollus et al. 1990).

Many seeds of new labour market conflicts may have thus been sowed during the recession. Will these developments undermine some elements of the labour relations largely based on consensus and negotiation? These labour relations have also paved way for the introduction of technical renewals, and made further incremental advances possible by learning and developing them at the shop-floor level.

1.2.4 Technology in National Development and Attitudes

Industrialization, Technology and Nationalism. Public opinion and attitudes about technology are important in creating the climate for technological development in the national context. Especially relevant are interest group attitudes and actions. In many countries, trade union opposition has been able to retard the introduction of technical innovations. Through political mechanisms, the prevailing ideological climate towards technological technical change also paves way for - or impedes - technology policies and other forms of public technology support.

In the Finnish history, industrial development has played a special role. It has been the other side of nationalism, the 'great project' of building up an independent and prosperous Finland. Thus, industrialization has not encountered much resistance, although political parties have had varying attitudes. All 'progressive' issues related to national development - higher education, strengthening national culture, creating strong industry - have been generally favoured. This is exemplified by old newspaper articles proudly announcing the establishment of a new - 'biggest in the world' - power station or paper mill (see e.g. Kuisma 1991, also Myllyntaus 1991).

This seems to be true with regard to the trade unions as well. Especially, while industrial development has been expanding, technical advance has not encountered notable criticism. The general belief in the positive effects of progress and technical development is also reflected in this attitude. The problems of employment and changing working conditions in the context of individual firms and plants have been compensated by the generally increased industrial employment.

This continued even during the 'informatization' period from the late 1970s onwards. Trade unions more or less supported the introduction of microelectronics. This is well exemplified by two books dating from 1979 documenting seminars held by the Central Labour Union, SAK and the other by the Metal Workers Union, Metalliliitto. Experts in technology and representatives of the two left-wing parties, speaking to trade union decision makers, all strongly supported technical advance. The only cause of concern was the 'taking care of the possible negative effects of automation' (SAK:n automaatioseminaari, 1979; Työ ja teknologia, 1979).

There are two special conditions prevailing in Finland and not in most of the older industrial countries, which may explain this:

First, the corporate three-partite system had been well-established by the time of the breakthrough of microelectronics', which made it easier to create formal negotiative procedures to deal with technical renewals. This is linked to the fact that the two left-wing parties were in the Government in the late 1970s, and thus the labour movement was, in fact, deeply involved in creating the new technology policies.

Second, the informatization of Finnish industry took place in a period when industry was expanding. Industrial employment only started to decrease, and modestly, from 1982 onwards. The introduction of microelectronics was thus not connected with decreasing employment as closely as in most of the older industrial countries. Subsequently, fears of unemployment were well muted by technical optimism. Hence, even quite comprehensive technological renewals, such as the transition to digitalized printing and the automation in the banking sector, have been carried through by negotiations with a minimum of conflict.

However, many problems may have been hidden during the period of economic growth. It seems that Finland is only in the 1990s facing the deindustrialization problems typical in Europe in the 1970s. While the introduction of microelectronics took place in the context of expanding industrial output and a growing service sector, the later developments have taken place in an environment of decreasing industry, decelerating growth of private services and a need to cut down the public sector. Simultaneously, the strength and unity of trade unions and the corporate negotiative system are disintegrating. This may pose new problems even for introducing technical innovations in the economy in the future.

Technical Change and Entrepreneurial Spirit. The nationalist emphasis can also be seen at the level of entrepreneurial activity. Kuisma (1991) observes the exceptional drive of Finnish entrepreneurs, especially in the years after the Second World War, to devote their time and resources to develop their businesses. This is best exemplified by the stories of many pioneers of technology, recalling the constant lack of money and the difficulties of acquiring even the simplest raw materials. However, raw materials and equipment had to be obtained somewhere: substitutes could be used, often equipment had to be made by oneself, sometimes even illegal methods to gain obtain the necessary materials had to be used (see e.g. Jauho 1992).

Kuisma highlights that business and profit related motivations were far from enough to explain the activities. The underlying driving forces were more general, and the personal ambitions were interwoven with the beliefs of national development and the rebuilding of the country after the destructions of the war. On an ideological level, the drive to build up an independent national state in the interwar years was continued during the reconstruction period after the Second World War. Kuisma concludes that if profitability had been the main criterion for industrial and technical development, no paper machines would have been built and most of the other technical advances would have remained undone as well. The activities of entrepreneurs and innovators in the postwar years were far from based on calculated comparative advantages and profit expectations.

Seen in the historical light, the current situation may perhaps also be blamed for lacking entrepreneurial spirit. Profit expectations do not suffice, especially in economically hard times. Did the years of easy finance and hasty stock market profits of the 1980s lead to a wrong image of the market economy?

1.2.5 Policy for Developing Technology

Origins of Technology Policy. Early Finnish industrial policy supported - already before the country's declaration of independence - especially investments that favoured the diffusion of new mining and agricultural techniques (Hjerppe 1990). The trend has continued, and technical novelties have been supported by industrial policy during the interwar period and after the Second World War.

However, the policy was mainly diffusion policy, supporting investing in new machinery and equipment. The main target was to raise price competitiveness and cut down production costs by introducing up-to-date production methods. Development of new products and research in private companies did receive much less public support.

On the other hand, the creation of the network of research institutes had begun quite early. The Helsinki Polytechnical School was established in 1872 and converted into the Helsinki University of Technology in 1908. Sectoral research institutes established by industry and industrial organizations also had a relatively early start. For example, the Central Laboratory, the research institute of the forest industries, was established soon after the declaration of independence. The Technical Research Centre of Finland, the VTT, was established by the state in 1941.

Up to the 1960s, technical colleges, universities and the VTT were the main agents of technology policy. Neither were any selective targets formulated, nor were any specific instruments to guide firms' research activities introduced. The foundations for a separate technology policy infrastructure were created in the 1960s: a technology unit was established at the Ministry of Trade and Industry and the Finnish National Fund for Research and Development, SITRA - a special credit institution to support technical research - was founded under the Bank of Finland. New policy instruments were also introduced, and product development loans, subsidies to companies and support for applied research in research institutes and universities were granted. However, in the 1960s and 1970s, technology policy did not, unlike industrial policy, follow any clear selectivity in allocating resources: support decisions were made by the Ministry on the basis of applications. A quantitative increase of resources was almost the only explicit target, and also the most important advance in the policy system in the 1970s.

Technology policy was criticized during the late 1970s in two respects. First, the system was blamed for bureaucracy, inefficiency and the lack of technical expertise. Second, new technical advances and the rapid diffusion of microelectronics raised the question of selectivity in targeting technical development.

A New Phase. The Technology Committee of 1980 became a turning point in technology policy. The Committee recommended a set of national technical development projects, which represented a new way of thinking in technology policy and a new level of state intervention in technical change as well. The recommended projects focused on semiconductors, information technology, automation in batch production and biotechnology in process industry. The target has clearly been to steer technical change towards microelectronics and information technology. Moreover, the traditionally dominating area of the forest industries is missing from the core of recommendations.

After the Committee, from the early to mid-1980s, the policy system developed rapidly. The most important institutional renewal was the establishment of the TEKES, the Technology Development Centre, based on the model of the Swedish STU, in 1983. The main task of the organization was to take over most of the technology activities from the Ministry of Trade and Industry - the finance and administration of research - and, above all, to create a more targeted and planned technology policy.

One of the first duties of the TEKES was to carry out the comprehensive national technology programme with the value of about FIM 190 million. A common denominator for all the twelve projects in the programme was high technology, especially information technology, which was through this programme, in fact, adopted as one of the main development targets of Finnish industry. Substantial selectivity was also introduced as an essential feature of technology policy.

Another new feature in the national technology programme was its way of organization. For the first time, the programme was consciously organized in the form of close cooperation between universities, research institutes and companies, thus even formally supporting the creation of user-producer networks.

Since the late 1980s, technology programmes have focused on generic technologies applicable in many areas, and technologies promoting the renewal of the forest industries and the metal and engineering industries, the nationally most important industries. The technology programme committee of 1989 singled out 11 areas of national technology programme development:

1. Advanced applications of electronic components;
2. Chemical synthesis technology;
3. Development of process systems;
4. New production technologies and computer-aided planning and production;
5. Advanced information processing systems;
6. Materials in manufacturing industry;
7. Information technology applications in products;
8. Telecommunication technology;
9. New pulp and paper technology;
10. Wood products and production;
11. Advanced construction, building materials and materials manufacturing.

Another typical feature of technology policy in the early 1990s is the increasing participation in international research organizations and projects. Finland has participated in the EC Eureka programme from the beginning. In December 1986, Finnish companies participated in 12 projects of a total of some 100 Eureka projects, while by the summer of 1992, the number had increased to the participation of Finns in 80 projects.

For the time being, there are not many comprehensive and systematic analyses of the impact of the various measures available in Finland. Single projects and programmes have been evaluated by the VTT, the TEKES and the Academy of Finland and by some research institutes. The OECD has also published its evaluations of the Finnish technology policy, and the assessments have generally been fairly favourable:

> The Examiners recognised ... that Finland has established an impressive collection of science and technology policies and now possesses a sizeable scientific and technological potential, both in terms of institutional diversification and resources. ...
>
> The vigour of the national research and development effort is impressive. The importance of science and technology in Finland's economic and social development should be acknowledged in the other sectors, so that adjustments can be made to remove obstacles and stimulate growth (OECD 1987).

Other Policies. Technical change has also received plenty of direct and indirect support through other public policies. Educational developments have been mentioned above: the supply of sufficient manpower through technical education has been among the central goals of educational planning. The strengthening of universities and research institutes has also been a main target of education and science policy.

Regional policy is another field in which developments related to information technology were taken seriously. This was based on the notion that the decline of old industries and agriculture in connection with the growth of new industries threatened to raise the problems of periphery onto a new level. This led to the development of regional technology policy, with, e.g., all provinces launching their own technology policy programmes in 1989. The underlying concept was to build up a new infrastructure for 'information society' (see e.g. Vuorinen 1990).

Activities in regional policy have led to new institutional arrangements as well. Technology parks were already mentioned: they are seen as an important means of developing and diffusing technology in their regional environments. In addition to technology centres, information technology centres which concentrate on training and supplying information technology services to firms in their environment. The centres are usually set up in close contact to - or as a department of - a local business or technical college. These organizations are supported both by education and regional policy. One of their aims has been to expand cooperation between public educational organizations and local small firms.

In the early 1990s, the regional policy activities have partly become a daily routine, although the most intensive drive to develop a regional information society has also slowed down. This is, on one hand, due to the recession, and on the other hand, to the relatively modest results. The production structures of peripheral areas have not been much modernized, and high technology firms have not, to any larger extent, settled into regions outside the clusters mentioned above. Has regional technology policy thus only succeeded in retarding the decrease of peripheral regions?

1.3 Technological Performance of the Finnish Innovation System

1.3.1 Research Efforts and the Finnish Manufacturing Industry in International Comparison

In the last three decades, the structure of the Finnish manufacturing industry has experienced important changes. Of the larger industrial sectors, the metal and engineering industries and the chemical industries have increased their share in total manufacturing output, while the shares of the forest industries and the textile and clothing industries have decreased (Table 1.6). However, within these large

Table 1.6. Manufacturing output by industry in 1960 - 1990

	Percentage shares at 1985 prices			
Industry	1960	1970	1980	1990
Food manufacturing	15.70	14.27	12.14	11.44
Textile, wearing apparel and leather	12.11	9.57	8.28	3.96
Total forest	24.33	23.42	19.81	18.10
- wood industry	9.77	7.34	6.15	4.61
- paper and pulp industry	14.56	16.08	13.66	13.49
Chemical industries	5.50	11.09	11.71	12.03
Metal and engineering industries	27.61	27.42	33.39	39.21
- basic metal industries	1.89	2.76	3.97	4.21
- manufacture of fabricated metal products, machinery and equipment	25.73	24.66	29.42	35.00
Other manufacturing	14.74	14.22	14.67	15.27

Source: ETLA

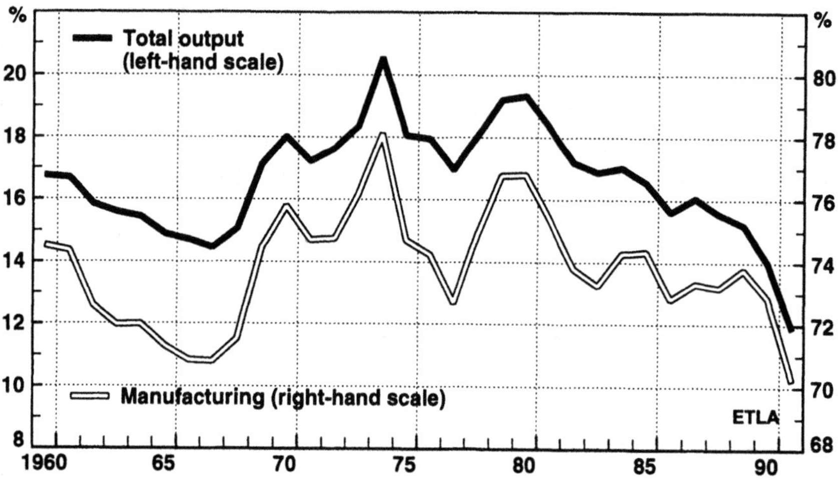

Fig. 1.3. Share of the open sector within total output and manufacturing. Source: ETLA.

sectors, the changes in the product and market structures have been even larger. Thus, for example, in the paper industry, there has been a remarkable shift towards moreupgraded products. Within the last 15 years, the share of the open sector in total output has decreased, partly because of the collapse of trade with the former Soviet Union, but Finland can still be regarded as a highly open economy (Figure 1.3). This means that Finnish products broadly compete in the world market. A decisively important part of international competitiveness is based on the technological performance of manufacturing.

Rapid Growth but a Medium Level of R&D Expenditures. Finnish industry started to intensively raise its technological level relatively late, but since the beginning developments have been fairly rapid. From the period from the 1950s to the 1960s, there is only scattered quantitative information on the trends of Finnish research intensity, and the figures are not fully comparable with the time series material which has been systematically gathered from the late 1960s onwards. For example, on the basis of a survey concerning 1956, it was estimated that total research expenditures (excluding research at universities) amounted to 0.35% of GDP in that year. Within manufacturing, the metal and engineering industries were already at that time the major spenders on research, with a 61% share of the total. In contrast, the forest industries had a 13% share while the chemical industries, still at the emerging industry stage in Finland, only spent 9% of the total manufacturing research expenditures (See Elfvengren 1958). According to another study, the share of technical research in GDP increased from 0.39% in 1962 to 0.49% in 1968. In the latter year, total research expenditures were estimated to be 0.8% of GDP; in Sweden the corresponding figure was already 1.6% (Nurmi 1970, p. 27-28).

In the late 1970s and 1980s, investments in R&D increased very rapidly in Finland. The annual growth rate of the volume of R&D was nearly 10% on average in the 1980s, or higher than in most of the other OECD countries. However, compared with the major spenders on R&D, the level of R&D investments is still fairly low. Instead, R&D expenditures relative to the level of GDP are in Finland on a comparable level with, e.g., Canada, Denmark and Norway (see Figure 1.4).

When looking separately at the R&D spending of the manufacturing industries, the international picture is quite similar to that on the national level: while manufacturing spends relatively more on research than the economy as a whole, the major spenders are the same countries (the U.S., the United Kingdom, Germany, Japan and France among the larger countries, and Sweden among the smaller ones, see Figure 1.5). In absolute terms, the industries which spend the largest amounts on R&D in Finland are, just as in most other industrialized countries, the metal products and engineering industries and the chemical industries. The former category of industries accounts for as much as two thirds of the total research expenditures of the manufacturing industries (Figure 1.6).

A more detailed analysis shows that there is considerable variation between subsectors of the different industries as to how much they spend on R&D. In some

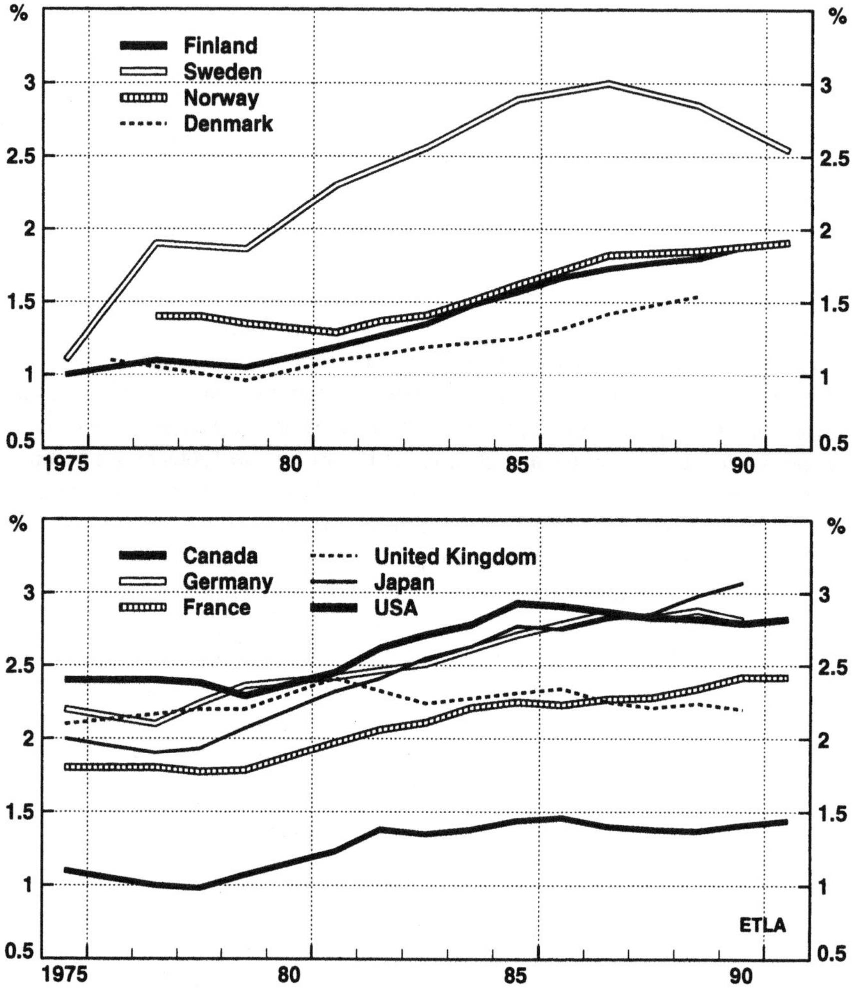

Fig. 1.4. R&D as a percentage of GDP in selected countries. Source: OECD.

fairly narrow product categories the share of R&D expenditures in value added was, for instance, in 1989 quite high: 24% for drugs, 55% for petroleum and coal products, 37% for computers and office machines, etc. (Table 1.7). Most of these sectors are regarded as genuine high-tech industries.

In contrast, the pulp and paper industry is quite a different example, which is especially important in the Finnish industrial structure. In international classifications, the industry is regarded as a low-tech industry, whereas in Finland the product category, as a whole, spent in 1989 only 2.2% of its value added on R&D - about half of the manufacturing average. Nevertheless, parts of the paper industry have distinct characteristics of a high-tech industry: without a high level of

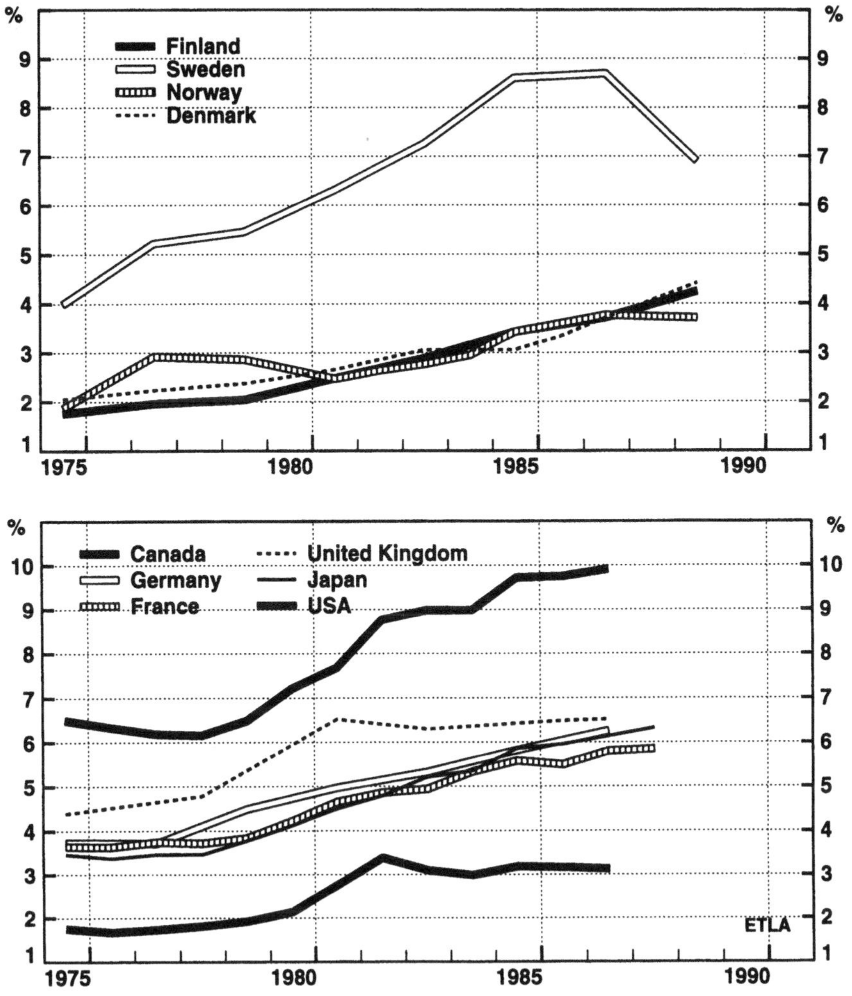

Fig. 1.5. R&D as a percentage of value added in manufacturing in selected countries. Source: OECD.

expertise and ability to use technologically complicated processes, the manufacture of high-quality papers would simply not be possible. A substantial part of the technologies used in the industry are created outside the sector, i.e. in the sectors producing machinery and equipment. The research efforts made in those sectors benefit the paper industry as well. In fact, the sector producing pulp and paper machinery is much more research-intensive: in 1989 it spent 9.5% of its value added on R&D.

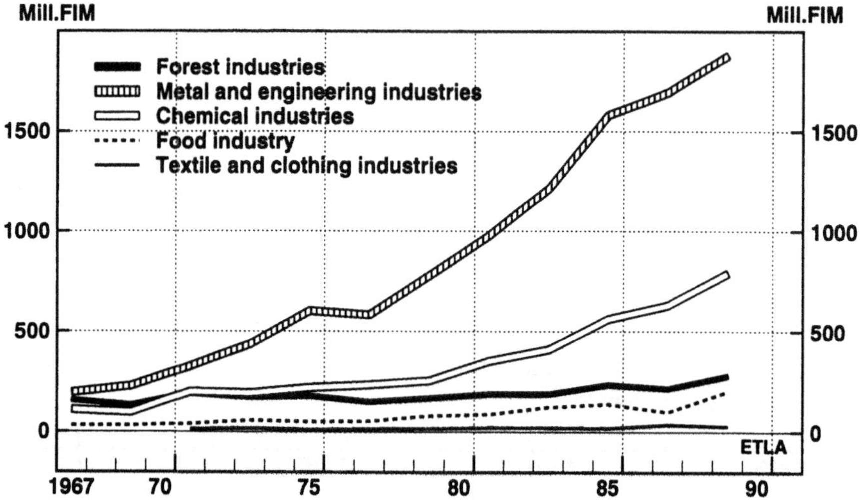

Figure 1.6. Volume of R&D in Finnish manufacturing industries (1985 prices). Source: Statistics Finland.

Table 1.7. Research and development expenditures of Finnish manufacturing firms in 1989, share in value added, per cent, selected product groups

Product group	%
Food	1.8
Clothing	0.1
Wood and wood products	0.5
Pulp, paper and paper products	2.2
Industrial chemicals	8.7
Drugs	24.2
Petroleum and coal products	54.9
Pulp and paper making machines	9.5
Computers and office machines	37.1
Instruments	14.7
Total manufacturing	4.7

Source: Statistics Finland

In addition, the pulp and paper industry funded research activities performed outside the firms themselves with an amount which was about one-fourth of the industry's "intramural" research expenditures; this is an exceptionally high share. Thus, the picture got by looking at the firms' own R&D expenditures alone is far from complete; in many cases, the diffusion and use of technologies developed by other firms or research units may be highly important.

It should also be remembered that the industrial research activities are highly concentrated on large companies. For example, in 1989 the ten largest manufacturing firms accounted for 54% of the total R&D expenditures of manufacturing, while their shares in total sales and exports were only about 44% and in the number of employees only 40% (Table 1.8). The degree of concentration increased very rapidly in the 1980s.

Table 1.8. Ten largest manufacturing companies: sales, exports, R&D and employment, share in total manufacturing, 1983 and 1989

	Sales		Exports		R&D		Employment	
Company	1983	1989	1983	1989	1983	1989	1983	1989
Nokia	3.7	8.9	3.3	5.7	14.4	22.2	4.4	8.7
Valmet	2.6	3.9	3.4	3.8	4.9	7.5	2.9	4.0
Rauma-Repola	3.1	4.7	5.2	4.5	2.6	3.3	3.4	4.0
Kone	2.1	2.8	1.6	1.3	1.8	3.1	2.4	3.8
Enso	3.0	4.2	5.2	7.0	1.3	1.5	2.8	3.8
Kymmene	2.4	4.4	3.0	7.4	4.6	1.2	3.0	3.5
Kemira	2.1	4.3	2.4	2.3	4.4	6.2	1.5	3.4
Outokumpu	2.0	4.6	4.5	6.8	7.6	4.3	1.9	3.3
Partek	1.1	2.9	0.4	0.8	2.5	3.5	1.2	2.8
Metsä-Serla	1.7	3.4	2.7	4.9	1.2	1.1	1.5	2.7
Total	23.9	44.2	31.6	44.4	45.3	53.9	24.9	39.9

Source: The large-firm data base of the Nordic Perspective Group

Structural Shift From Fixed Investment to Research Investments. In most industrialized countries, the manufacturing industries have increased their R&D expenditures much more rapidly than investments in fixed capital over the past 15 or 20 years. This structural shift has been remarkable in countries like the U.S. and the Nordic countries. Despite the exceptionally rapid shift in emphasis of investments in Finland (see Figure 1.7), the relative level of intangible investments is still far behind Sweden, as noted above. The much faster growth in research expenditures than the growth of fixed investments can be clearly seen also when looking at the growth rate differentials of the main subsectors of manufacturing in the 1980s (Table 1.9).

In Finland, the investment ratio has been internationally high in the postwar period. In the 1980s, gross fixed capital formation amounted to about 25% of GDP, on average, which was almost 5 percentage points higher than the EC average and about 3.5 percentage points higher than the average for small European OECD countries. This high ratio is partly explained by the rather heavy investments in the housing sector and in agriculture, but also within manufacturing the investment ratio has remained at a relatively high level, although there has been a long-run downward trend (see Figure 1.8).

Among the larger industrial sectors in Finland, the wood, pulp and paper industries are the most capital-intensive ones: in 1989 they accounted for 44% of the fixed capital investments of total manufacturing, and they spent 12.5% of their turnover on these investments, whereas the manufacturing average was 7%. At the same time, these industries' share in the total intangible investments (including R&D, training, marketing, software, etc.) of manufacturing was only 15.5%; they spent 1.7% of their turnover on intangible investments, while the manufacturing average was 2.7%. These figures can be compared with the corresponding figures for the industries producing machinery and equipment, which are much less capital-intensive but more research-intensive. The share of these industries in fixed

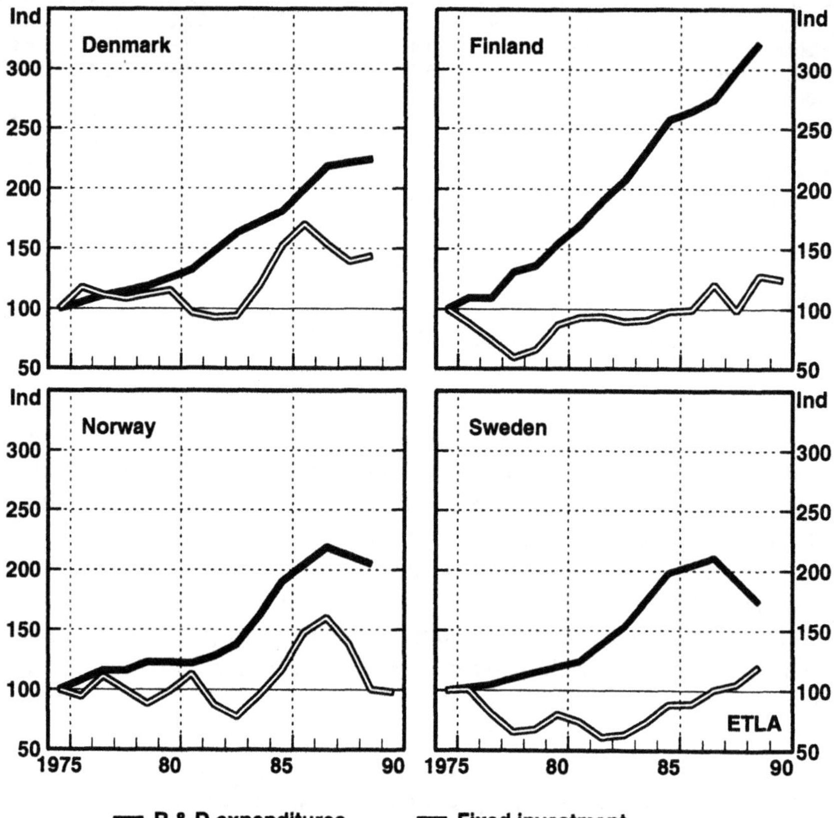

Fig. 1.7. Volume of R&D expenditure and fixed investment in manufacturing in the Nordic countries (1975=100). Sources: Statistics Finland; OECD.

Table 1.9. Manufacturing fixed investment and R&D expenditures by industry in 1980 and 1990

	Fixed Investment			R&D Expenditure		
	Share in total manufacturing %		Average annual change in volume %	Share in total manufacturing %		Average annual change in volume %
	1980	1990	1980-1990	1980	1990	1980-1990
Food manufacturing	13.1	10.6	1.4	5.1	7.2	13.7
Textile, wearing apparel and leather	4.7	1.9	-5.1	1.1	0.7	6.8
Forest	32.1	38.2	4.5	11.0	8.5	10.0
Chemical	12.3	12.2	3.8	19.0	23.7	15.1
Metal and engineering	26.4	24.6	3.8	61.3	55.6	10.1
Other manufacturing industries	11.4	12.5	5.5	2.5	4.3	16.2
Total manufacturing	100.0	100.0	3.7	100.0	100.0	11.3

Source: Statistics Finland

Note: R&D expenditures have been deflated by the implicit GDP deflator.

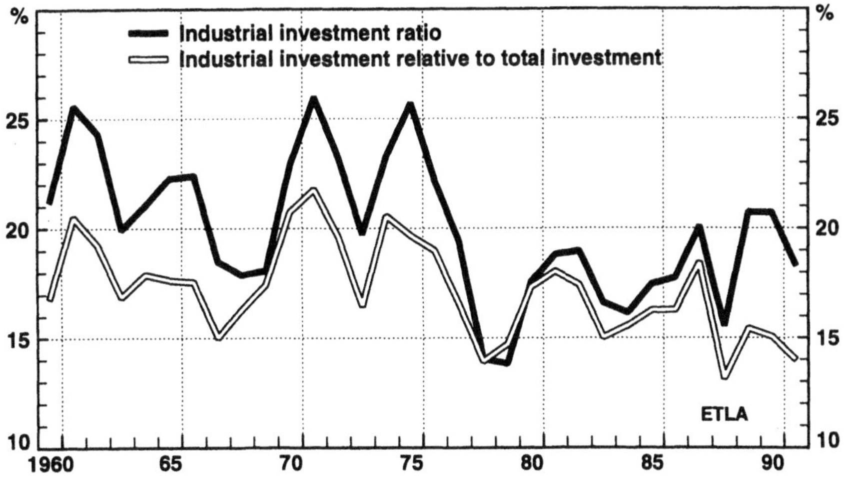

Fig. 1.8. Industrial investment ratio and share of industrial investment within total investment. Source: ETLA.

capital investments of total manufacturing was only 6.5%, whereas their share in the intangible investments of total manufacturing was as much as 14.5%. As mentioned above, part of the research results of this sector are transferred to the users of the machinery and equipment produced by this sector.

1.3.2 Industrial Growth and Results of Research Efforts: Measures of Technological Advance

While it is clear that the growth of Finnish R&D investments has been internationally quite rapid in the 1980s, the results of these investments are not easy to measure. The next section deals first with productivity developments, which are, in general, thought to be related to the changes in research efforts. Another way of trying to find out the results of increasing research investments is to look at the trends in the patenting activities and trade in high-tech products.

Productivity Developments. By international standards, Finnish industrial growth has been rapid in the postwar period. This growth has been almost continuous, two notable exceptions being the mid-1970s and the early 1990s. Within this period, however, there has been a marked change in the character of growth: the earlier years were characterised by extensive growth based mainly on an increased use of the basic inputs, labour and capital. In contrast, since the early 1970s, growth has been of a more intensive kind, based on factors such as a more efficient use of inputs, better organization and technological advance. In other words, much more than before, the growth in total factor productivity (TFP) has contributed to the growth of industrial output.

Table 1.10. Growth of manufacturing output in Finland and contributions of labour and capital input and total factor productivity (TFP), average annual changes, per cent

1960-1973	Output growth	Contribution of labour	Contribution of capital	Contribution of TFP
Food manufacturing	5.1	0.8	1,8	2.5
Textile, wearing apparel and leather industries	3.9	-0.8	1.0	3.7
Wood industry	4.1	-0.6	1.1	3.6
Furniture industry	7.0	1.0	2.7	3.3
Pulp and paper industry	7.0	0.7	2.2	4.1
Printing and publishing	4.3	1.6	2.2	0.5
Chemical industries	13.2	2.4	5.6	5.2
Non-metallic mineral products	10.4	1.2	3.2	6.0
Basic metal industry	10.4	2.4	3.5	4.5
Metal products and engineering industries	6.7	1.4	1.7	3.6
Other manufacturing	7.0	-1.2	1.3	6.9
Total manufacturing	6.6	0.7	2.2	3.7

1973-1990	Output growth	Contribution of labour	Contribution of capital	Contribution of TFP
Food manufacturing	2.2	-0.9	0.9	2.2
Textile, wearing apparel and leather industries	-1.6	-3.9	0.1	2.2
Wood industry	0.5	-2.4	0.6	2.3
Furniture industry	2.7	-0.9	0.9	2.7
Pulp and paper industry	2.3	-1.1	1.2	2.2
Printing and publishing	4.1	0.3	2.6	1.2
Chemical industries	3.2	-0.5	1.6	2.1
Non-metallic mineral products	2.7	-0.7	1.2	2.2
Basic metal industry	5.2	-0.4	0.9	4.7
Metal products and engineering industries	4.9	-0.3	1.4	3.8
Other manufacturing	3.1	-1.2	1.3	3.0
Total manufacturing	3.1	-0.9	1.1	2.9

Source: National accounts and calculations made by ETLA

Indeed, the contribution of TFP to output growth in manufacturing was about 56%, on average, from 1960 to 1973, whereas it increased to as much as 94% in the post-1973 period (Table 1.10). However, there are large variations both between industries in both periods and between periods for each industry. For instance, the average growth contribution of TFP in the chemical industries only increased from 39 to 66% between the periods, whereas in the wood industry it grew from 88 to 460%. At the same time, there was a marked slowdown in the average annual growth rates of output for both industries, just like in total manufacturing.

The rate of growth of both total factor productivity and labour productivity in the business sector as a whole, was in the 1980s clearly higher in Finland than in most of the other OECD countries (Table 1.11). In addition, the slowdown of productivity growth as against the 1960s and the early 1970s has not been as strong as generally in other industrialized countries. The factors behind this slowdown are, however, probably partly the same for Finland as for the industrialized countries in general: the effect of large changes in energy and raw material prices, the slowdown of capital accumulation, lower expenditure on R&D, demographic change affecting the average skill level of the labour force, and reduced opportunities for "catching up" since the 1950s and 1960s (OECD Economic Outlook).

Studies of the effects of firms' own R&D activities on total factor productivity have, in general, produced the conclusion that, especially before the economic crises of the 1970s, the returns on R&D have been very high, often in the range of 20-80%. Most of these results, however, have concerned the U.S.A. and other large countries. Similar analyses based on Finnish, Swedish and Norwegian data have indicated that the returns on firms' own R&D are perhaps not as high in these countries as has been thought on the basis of previous research (see Vuori 1986 and 1991).

Table 1.11. Productivity in the business sector in selected countries, percentage changes at annual rate

Country	Total factor productivity			Labour productivity		
	1963-73	1973-79	1979-90	1963-73	1973-79	1979-90
United States	1.6	-0.4	0.3	2.2	0	0.7
Japan	5.9	1.4	2.0	8.6	2.9	3.0
Germany	2.7	1.8	0.8	4.6	3.1	1.6
United Kingdom	2.3	0.6	1.6	3.6	1.6	2.1
Denmark	2.8	1.2	1.3	4.3	2.6	2.1
Sweden	2.7	0.3	0.9	4.1	1.5	1.7
Finland	3.2	1.5	2.5	4.9	3.2	3.6
OECD Europe	3.3	1.4	1.3	5.0	2.7	2.0
OECD	2.8	0.5	0.9	4.1	1.4	1.5

Source: OECD Economic Outlook

It has thus been argued that the role of firms' own R&D and that of technology diffusion may be different in small open economies, like Finland, from the situation in larger and earlier industrialized countries. Diffusion seems hence to be extremely important for smaller countries, but it can be achieved only by means of sufficient own capabilities of firms, which are partly built on their own research activities. The decisive role of diffusion could explain the fact that very strong empirical evidence of the positive impact of firms' own research on their productivity has so far not been found for the Nordic countries (Vuori 1992).

Patenting and Trade in High-Tech Products. As for many other industrialized countries, the United States is the most important foreign country where Finnish applicants apply for patents. In the 1980s, there was a rapid increase in Finnish patenting activities; the average annual growth rate for patents granted in the U.S. was 8.5% during the decade. This rate can be compared with the growth rate of domestic patent applications: only 4.5% a year, on average, (see Table 1.12). The growing intensity of patenting activities can be seen as an indication of increased research activities and of more patentable research results, but also possibly of changes in patenting behaviour, as a result of, for example, keener international competition.

Also in terms of being able to produce exportable high-tech products, the performance of the Finnish manufacturing industry has improved considerably in the 1980s. While the share of high-tech products in total merchandise exports is still low by international standards, it has grown steadily, and in the last few years very rapidly. The growth in high-tech imports has been much slower, and thus the relative size of the high-tech trade deficit has decreased. In 1990, the share of high-tech exports in total Finnish exports was higher than in Denmark and Norway, but still clearly lower than in Sweden (Table 1.13). A trade deficit exists in almost all

Table 1.12. Patenting activities of Finnish applicants, 1980 to 1990.

Year	Patents applied in Finland by Finnish applicants	Patents granted in the U.S. to Finnish applicants
1980	1354	140
1981	1419	164
1982	1633	157
1983	1713	136
1984	1769	181
1985	1719	227
1986	1751	222
1987	1851	293
1988	1977	241
1989	1944	271
1990	2068	320
Growth, average 1980-1990	4.5 %	8.5 %

Sources: Valkonen and Lovio 1991 and Statistics Finland.

Table 1.13. The share of high-tech products in foreign trade in some OECD countries, %

	1981		1983		1985		1987		1990	
Country	I	E	I	E	I	E	I	E	I	E
Finland	10.1	3.7	11.5	5.1	13.1	6.2	16.2	8.6	15.2	10.8
Sweden	11.5	12.0	14.4	13.0	14.4	13.0	16.3	14.1	16.6	15.2
Norway	10.1	4.4	11.2	4.1	13.1	4.1	13.5	6.0	13.6	5.0
Denmark	10.0	8.2	11.1	8.0	12.1	10.0	13.8	10.0	14.5	9.0
Austria	9.6	8.3	10.6	10.9	11.9	12.1	13.1	13.9	14.9	15.7
Germany	10.0	13.8	11.0	14.9	12.7	15.3	14.3	16.0	14.9	15.7
UK	12.7	16.6	14.8	17.2	16.4	19.2	17.1	19.0	17.5	20.1
France	11.3	12.3	12.4	13.5	13.5	15.1	15.8	16.2	16.1	16.2
Japan	4.3	16.1	5.6	18.9	6.5	0.3	8.1	24.5	9.7	26.6
USA	9.1	17.4	12.1	22.5	13.6	24.5	15.7	25.1	17.7	25.4

Source: Statistics Finland

Note: I=Imports, E=Exports.

product categories, the most important exception being telecommunication equipment: in this product category exports in 1990 exceeded imports by as much as 86% (Teknologian soveltaminen ja siirto 1990, Koulutus ja tutkimus 1992:2).

To summarize, Finland has been able to raise her technological level considerably in the last two or three decades. Research efforts have grown more rapidly than in most of the other OECD countries. Technology diffusion has also been extremely important, as Finland has been able to adopt technologies created elsewhere and to adapt them to her own needs. Increased investments in research, education and technologically advanced machinery have produced many favourable results: increased productivity, more high-tech exports and more international patents. However, the levels of the technological leaders have not yet been reached, and probably will never be reached - except for a few fairly narrow niches - because of the relatively limited resources of a small country like Finland. The results of the extensive intangible investments in the 1980s have, to some extent, not yet been seen. Was the allocation of resources in that decade successful? The relative position of Finland in the international technology race of the 1990s hence continues to be an open question.

1.4 Concluding Remarks

Tentatively, the specific historical features of the Finnish innovation system could be summarized in five factors:

1. Technical development started, as a dimension of basic industrialization, both relatively *late and slowly*, but in a fashion that could accumulate a substantial and solid competence base - also in institutional terms - for further development.
2. The slow development was linked to the exceptional *nationalism* in the development model. The role of foreign capital and technology transfer in industrialization was unimportant, and technical knowledge was acquired mainly through nationals' activities.
3. On one hand, the *narrow basis of industrial structure* limited the scope of technical development, but could, on the other hand, help in creating a closely working development block based on long experience of user-producer cooperation. This co-operation expanded the innovative activities up and down the production chain and fostered the creation of a network of supporting institutions.
4. *Ideological and value related* factors created a favourable climate for steady technical advance. Strong drive for independent national and economic development, a distinct belief in the value of education and knowledge in combination with the devoted work values - perhaps inherited from the very recent agrarian past - supported industrialization and technical development both as generally accepted national aims and in practice by mobilizing the existing skills into practical development work.
5. Later on, the creation of a *corporate welfare state* that integrated trade unions to a three partite negotiative system and a largely consensus based idea of social development facilitated the introduction of new technologies, the acceptance of technology based 'structural change' as an economic policy target and the creation of a relatively straightforward and rational technology support system.

In the international comparison, the notably rapid technical advance from the early 1970s onwards sprang up from this kind of a historical background. Until then industrial growth had still been characterised by extensive growth based mainly on an increased use of the basic inputs, labour and capital. Since the early 1970s, growth has been more intensive, based on factors like a more efficient use of inputs, better organization and technological advance.

During this period, Finnish industry narrowed remarkably its technology gap to older industrialized countries. This was indicated by, e.g., the exceptionally high annual growth in the volume of R&D, nearly 10%, on average, in the 1980s, which was much above the growth rate of fixed investments. The R&D expansion was related to a structural change, on one hand, towards high-tech industries and, on the other hand, integrating high-tech components into the basic industrial complexes. The Finnish innovation system has thus also shifted the main focus from the diffusion of existing technologies towards more research-based development.

There are inherent contradictions in the effects of growing research investments. A paradox is that while R&D investments have accelerated rapidly since the early 1970s, both the average annual manufacturing output and productivity growth rates have slowed down markedly - although less than in most other industrialized countries. Moreover, any strong evidence of the positive impact of firms' own research on their productivity has so far been found, neither in Finland nor the other Nordic countries.

The slowdown of productivity growth could be explained by the sharp changes in energy and raw material prices, a deceleration in capital accumulation, changes in the industrial structure and perhaps weakened prospects for a further decline of the technology gap. However, do these explanations suffice, and what have really been the effects of growing R&D investments? Do they rather reflect the growing difficulties in further technical development than its success? Is the main role of firms' own R&D to facilitate the implementation and further incremental development of existing innovations and thus, in the first hand, to help accelerating diffusion?

However, in the 1980s, the share of high-tech products in total merchandise exports has grown, even very rapidly during the last few years. In 1990, the share of high-tech exports in total Finnish exports was higher than in Denmark and Norway. In telecommunication, the exports of 1990 exceeded imports by 86%, although a trade deficit actually existed in almost all other high-tech product categories.

1.4.1 The End of an Era?

In the early 1990s, the basic set-up is facing radical changes. The rapid growth in the 1980s has ended in a deep recession and led to an overall confusion. The forest industry sector seems to be, again, the cornerstone of the Finnish economy while the post-industrial information society has not risen from the ashes of the old industries.

Many of the original background factors of the Finnish model of progress - especially the ideological dimensions - seem to be eroding as well: There were no national aspirations behind the investors' drive to the overheated stock markets in the 1980s; the consensus-based system is largely questioned, and even the existing institutional infrastructure of a welfare state - including the basic principles of education - are questioned; and it is often argued that only traces of the old work values are left.

However, the more physical features of the model are also changing. As a result of internationalization, the national innovation system is opening from both ends. On the one hand, foreign capital has gradually begun to gain better access inside the country, even in some technologically quite central firms. On the other hand, Finnish firms have rapidly loosened their domestic ties through international operations. This is well illustrated e.g. by the relocation of some Nokia department headquarters abroad. While operating in an international context, the weight of national considerations further diminishes in the corporate strategies and decision making.

From a technological point of view, there is not much to gain from narrowing the technology gap any longer. The gap, of course, still exists, but the advantages from filling it will decrease. For example, in many cases, the technical and financial demands are growing rapidly also for companies pursuing a followers' strategy.

Further advance and increasing internal complexity of high-tech products is changing the nature of technical competition. Entry barriers and development costs will be higher, and there are likely fewer basic technical development paths to choose from. The direction of technological change may become, to a growing extent, predecided in the global centres of technology. Further on, the participation in development may - because of high development costs and entry to vital knowledge - also become more dependent on participating in international development cooperation. This will make it still harder for both small firms and small countries to stay on the edge of technical change.

In this situation, the whole national innovation system may have to be redefined. What will be the role of the national component, when most of the advances and operations are decided internationally, and even the substantial development is bound to international linkages? In the context of integrating Europe, firms do - by definition - even generally execute their transactions more loosely under less national limitations.

The same opening is true in respect of other components of the innovation systems. For example, firms are more open to use labour force from anywhere. The national pool of competence will thus face much more competition from two directions: it will become easier to recruit people from abroad and it will also be easier to relocate operations in other European countries. Similarly, it will be easier for technology intensive firms to get information and expertise from elsewhere in Europe, or relocate operations in the sources of knowledge.

However, the reality may not be that 'loose'. Both cultural issues and close interaction between users and producers of technology and other networking linkages make drastic relocations difficult for high-tech firms.

Another possible outcome from the development towards a more global world with less national limitations may be a world of local and regional milieus, based on intense interaction, tight linkages and dynamic development - milieus interacting and co-operating more with similar environments in other European countries than with other regions in the home country. But even such a development model implies radical changes in national policies and development targets.

We can hence summarize: in the last two or three decades, Finland has considerably raised her technological level. Increased investments in research, education and technologically advanced machinery have produced many favourable results: increased productivity, more high-tech exports and more international patents.

However, many results of the extensive intangible investments in the 1980s still remain to be seen. Was the allocation of resources successful? Were the target sectors the right ones: not much is left of the expressive growth of electronics. Telecommunication and consumer electronics remain in domestic hands, while foreign capital has taken over production within most of the other subsectors of electronics.

Moreover, what will be the future role of the forest sector? Will the country be retarded back into a producer of basic, raw-material based paper products? Or, could the forest industry sector once more - in analogy to the electronics sector - induce new dynamic areas of developing technology? What might they be: innovations related to biological or environmentally friendly paper products, or even other sectors of environmental technology, e.g. water protection and cleaning technologies?

The position and prospects of Finland and Finnish firms in the international technology race in the 1990s hence continue to be a widely open question: there are too many open - and new - parametres affecting technological competition in the near future.

References

Alasoini, T. (1990), Tuotannolliset rationalisoinnit ja teollisuuden työvoiman käyttötapojen muutos. Helsinki: työministeriö, työpoliittinen tutkimus 5 (in Finnish).

Asplund, R. (1991), Education, human capital and economic growth. Helsinki: ETLA (The Research Institute of the Finnish Economy), Series B No. 71.

Autio, E., M.M. Kaila, R. Kanerva and I. Kauranen (1989), Uudet teknologiayritykset. Helsinki: SITRA, No. 101 (in Finnish).

Edquist, C. and B.-Å. Lundvall (1989), Comparing Small Nordic Systems of Innovation. Linköping/Aalborg: mimeo.

Ekola, J., P. Vuorinen and P. Kämäräinen (1991), Ammatillisen koulutuksen uudistaminen 1980-luvulla. Selvitys uudistuksen toteutumisesta ja toteutusympäristöstä. Helsinki: Ammattikasvatushallitus (in Finnish).

Elfvengren, E. (1958), Kostnaderna för teknisk, naturvetenskaplig och ekonomisk forskning i Finland. Helsinki: Taloudellinen Tutkimuskeskus, Sarja B:10 (in Swedish with Finnish and short English summary).

Hjerppe, R. (1990), Kasvun vuosisata. Helsinki: Valtion painatuskeskus (in Finnish).

Johnson, B. (1988), Institutional learning and national systems of innovation. Aalborg: mimeo.

Kajaste, I., S. Parviainen and P. Ylä-Anttila (1992), Foreign direct investment of Finland in the 1990s. Helsinki: VATT, Government Insitute for Economic Research.

Kauranen, I., M. Takala, E. Autio and M.M. Kaila (1991), The first three years. A study on the first phases of a new science park and of its tenants. Espoo: Publications of Otaniemi Science Park. Publication Number 2.

Komiteanmietintö 1988:28, Väestön koulutus. Helsinki: Valtion painatuskeskus (in Finnish).

Kuisma, M. (1991), Nationalismi, industrialismi, Suomi. Helsinki: mimeo (in Finnish).

Kuitunen, T. (1991), Outlook for Manufacturing Industries 1991 - 1995, in: The Finnish Economy 1991/4. Helsinki: ETLA (The Research Institute of the Finnish Economy).

Lemola, T., T. Loikkanen, R. Lovio, R. Miettinen and P. Vuorinen (1990), Teknologiatutkimuksen kehittäminen Suomessa. Helsinki: TEKES. TEKES julkaisu 24/1990 (in Finnish).

Lemola, T. and P. Vuorinen (1988), "Finländsk teknikpolitik förnyas", in: Glimell, H. (ed.), Industriförnyelse i Norden. 80-talets programsatsningar på mikroelektronik. Århus: Forlaget Samfundsoekonomi og planlaegning.

Lilja, K., K. Räsänen and R. Tainio (1992), "Sectoral foundations of competitive advantage: the case of Finnish forest corporations." To be published in Whitney, R. (ed.), European Business Systems in a comparative perspective. London: Sage.

Lindell, M. (1989), Strategiutveckling i produktinnovativa företag. Helsinki: Swedish School of Economics and Business Administration. Research Reports 22 (in Swedish with an English summary).

Lovio, R. (1989), Suomalainen menestystarina? Tietoteollisen verkostotalouden läpimurto. Helsinki: Hanki ja jää (in Finnish).

Lovio, R. (1992), "The influence of foreign companies on the birth and development of the Finnish electronics industry", in: Vuori and Ylä-Anttila (eds.)(1992).

Lundvall, B.-Å. (1992), "Introduction", in: Lundvall, B.-Å. (ed.), National systems of innovation - Toward a theory of innovation and interactive learning. London: Pinter Publishers.

Lundvall, B.-Å. (1989), NIPS - a project on the Nordic Innovation and Production Systems. Aalborg: mimeo.

Myllyntaus, T. (1991) Electrifying Finland. The transfer of a new technology into a late industrialising economy. Helsinki and London: Macmillan in association vith ETLA.

Myllyntaus, T. (1992), "Technology transfer and the contextual filter in the Finnish setting. Transfer channels and mechanisms in a historical perspective", in: Vuori and Ylä-Anttila (eds.)(1992).

Nordisk Industrifond (1991), Nordisk FoU-statistikk for 1989 og statsbudsjettanlyse 1991, Oslo: Nordisk Industrifond (in Norwegian).

Nurmi, M. (1970), Tekninen tutkimus Suomessa vuosina 1962 ja 1965-68. Helsinki: Taloudellinen Tutkimuskeskus, Sarja B 17 (in Finnish with short English summary).

OECD Economic Outlook 50, December 1991. Paris: OECD.

OECD (1987), Reviews of National Science and Technology Policy. Finland. Paris: OECD.

Ollus, M., R. Lovio, J. Mieskonen, P. Vuorinen, J. Karko, S. Vuori and P. Ylä-Anttila (1990), Joustava tuotanto ja verkostotalous. Helsinki: SITRA, No. 109 (in Finnish).

Piironen, P. (1992), Kaksi esimerkkiä yhtiöittämisestä. Helsinki: Sisäasiainministeriö, kunta- ja aluekehittämisosasto (in Finnish).

Raumolin, J. (1990), The impact of technological change on rural and regional forestry in Finland. Helsinki: ETLA (The Research Institute of the Finnish Economy), Discussion Papers No. 343.

Ray, G.F. (1991), "Innovations diffused - a random walk in history", in: Vuori and Ylä-Anttila (eds.)(1992).

SAK:n automaatioseminaari (1979). Turenki: SAK (in Finnish).

Scott, A. J. (1988), New industrial spaces. Flexible production and regional development in North America and Western Europe. London: Pion.

Soete, L. (1988), "Technical change and international implications for small countries", in: Freeman, C. and B.-Å. Lundvall (eds.), Small countries facing the technological revolution. London and New York: Pinter Publishers.

Statistical reports of the Nordic countries No. 56 (1991), Stockholm: Nordic Statistical Secretariat.

Technology and the economy (1992), The key relationships, The Technology/Economy Programme. Paris: OECD.

Technology in a changing world (1991), The Technology/Economy Programme. Paris: OECD.

Teknologian soveltaminen ja siirto 1990 (1992). Helsinki: Tilastokeskus (Central Statistical Office of Finland), Koulutus ja tutkimus 1992:2.

Tisdell, C. (1982), Science and technology policy priorities of governments. London: Chapman & Hall.

van Tulder, R. (1989), "Studies of small industrial countries and economic and technological development", in: van Tulder, R. (ed.), Small industrial countries and economic and technological development. The Hague: Netherlands Organisation for Technology Assessment.

Työ ja teknologia (1979). Tampere: Metallityöväen liitto, r.y (in Finnish).

Valkonen, P. and R. Lovio (1991), Suomalaiset patentoijat Yhdysvalloissa 1980-luvulla: Menestyksen suhteellisuus. ETLA (The Research Institute of the Finnish Economy), Discussion Papers No. 383 (in Finnish).

Vuori, S. (1986), Returns to R&D in Finnish and Swedish manufacturing industries. ETLA (The Research Institute of the Finnish Economy), Discussion Papers No. 203.

Vuori, S. (1991), Returns to R&D in Nordic manufacturing industries, 1964 to 1983. ETLA (The Research Institute of the Finnish Economy), Discussion Papers No. 357.

Vuori, S. (1992), "R&D, technology diffusion and productivity performance in Finnish manufacturing industries", in: Vuori and Ylä-Anttila (eds.)(1992).

Vuori, S. and P. Ylä-Anttila (eds.)(1992), Mastering technology diffusion - The Finnish experience". Helsinki: ETLA (The Research Institue of the Finnish Economy), Series B No. 82.

Vuorinen, P. (1990), "Teknologiaverkostot ja yritysten toimintaympäristö - katsaus yrityksen ja alueellisen toimntaympäristön vuorovaikutukseen", in: Lemola, T., T. Loikkanen, R. Lovio, R. Miettinen and P. Vuorinen: Teknologiatutkimuksen näkökulmia ja tuloksia. Helsinki: TEKES. TEKES julkaisu 25/1990 (in Finnish).

Vuorinen, P.,T. Tikka and R. Lovio (1989), Suomen teknologiakeskukset. Helsinki: Sisäasiainminiesteriö, aluepolilittisia tutkimuksia ja selvityksiä 1989:5 (in Finnish).

Walsh, V. (1988), "Technology and the competitiveness of small countries: A review", in: Freeman, C. and B.-Å. Lundvall (eds.), Small countries facing the technological revolution, London and New York: Pinter Publishers.

2 Pecularities of Social and Technological Change in the Finnish Society

Erik Allardt
University of Helsinki and Åbo Akademi University, Finland

In the prosperous days of the second half of the 1980's it was in Finland not uncommon to ask whether Finland is an exceptionally modern country. The question was often attributed to foreign observers although it was usually triggered of by Finns and Finnish national ambitions. Yet, the idea of Finland being exceptionally modern was not only a reflection of Finnish ethnocentrism but it was also based on some widespread popular assumptions with roots in the Enlightenment and Western liberalism. The first is the idea that there exists a unique but central model of development towards which all countries move or wish to move. The second is the assumption that modernization can be assessed by some simple indicators of economic development. The third is that modernization as being the most important thing in the world is something all individuals in their right mind should aspire to.

All three assumptions are difficult if not altogether impossible to sustain. It seems reasonable to assume that there are alternative paths to development. Furthermore, modernization is a multidimensional phenomenon whereby the developments on different dimensions have to be accounted for. Third, from a human point of view its seems more important to give priority to problems such as order versus disorder, peace versus violence, equality versus inequality etc. than to just modernization. At any rate, in analyzing and describing the roots and consequences of technological change a whole array of circumstances outside the economic and technological realms of life have to be taken into account.

2.1 The Multidimensionality of Modernization

Processes of modernization occur both in politics, the social life and economy. They are all dependent on and influence each other. It is not reasonable to postulate that the economic and technological development are isolated phenomena or even that they can be treated and described without references to their political and social connections. Here particularly three aspects of modernization will be emphasized: *political mobilization*, *social differentiation* and *technological development*. A basic assumption in this paper is that an important part of the economic problems of the Finnish society is due to institutional structures. The famous political scientist Karl Deutsch (1961, 494-495) coined the term social mobilization to denote the political aspects of modernization. He defined social mobilization as "the process in which major clusters of old social, economic and psychological commitments are eroded and broken and people become available to new patterns of socialization and behavior". Many occurrences such as the response to the mass media, change of residence, urbanization, and literacy can be taken as indicators of social mobilization. Here the focus will be on political developments, whereby it seems feasible to speak directly about political instead of social mobilization. As a simple but crucial indicator of political mobilization voting participation in the general parliamentary elections will be used in the forthcoming text.

In social life one of the most important aspects of modernization is caught by the term social differentiation. Modern societies are highly differentiated and contain a vast amount of social roles and institutional structures. Recruitment to these roles is as a rule not ascribed in advance by kinship, territory, caste, status at birth etc. As one leading theoretician of modernization, S.N. Eisenstadt (1966, 3) once concluded "Perhaps the most important aspect of this differentiation and specialization of roles in all major institutional spheres is the separation between the different roles held by an individual - especially among the occupational and political roles, and between them and the family and kinship roles". Historically the latter type of separation of roles came first. Modernization meant that people were less and less ascribed to their occupational roles on the basis of their family background and kinship. In a later phase of modernization there was a separation of an individual's occupational roles on the one hand, and his political and cultural roles on the other. For instance, a given occupation does not automatically entail the incumbency of a particular role in politics. Furthermore, within each institutional realm (in politics, in the economy, in the sphere of social organizations, in territorial activities) there developed distinctive units that were organized around the goals specific to each realm. The different institutional realms are not in a modern society fused into a vast network in which family, occupational, territorial and political roles are clearly tied to each other.

It is often simply and flatly assumed that social differentiation has proceeded very far in the modern society, and that roles from different societal realms are not fused and strongly dependent on each other. This is an assumption which at least as regards the Finnish society can be seriously questioned. Finland was quite recently an agrarian society with an agrarian type of politics in which family, occupational, territorial and political roles were tied together. Despite dramatic changes in the political structure there are still many signs of an agrarian type of politics with a fusion of family, occupational, territorial and political interests. There is also plenty of evidence about instances in which political party membership has influenced the allocation of positions and jobs in other spheres of life. Furthermore, and perhaps most significantly there are signs of strong, almost rigid connections between occupational roles and political standpoints. Special-interest organizations have arisen in most walks of life. They are usually based on occupational roles but the interest organizations bargain not only for wages but also for political power, and they strongly influence the capacity of the society to adopt new technologies and innovations. Such a pattern is typical for what is labeled a corporatist society.

A crucial feature of modernization in the economic sphere is a rapid pace of technological development. Since technology may also have detrimental effects, the development of technology creates many moral and human problems. Yet, there has in all so-called modern societies occurred a development based on the systematic application of knowledge strongly fostered by preferences for a Newtonian type of science. This is also reflected in the fact that the growth of research expenditures has been regarded as one of the best indicators of modernization. It is well known that there was a rapid growth of the investments in research and development (R & D) in Finland in the 1980's, and that the country's R & D as a per cent of GDP in this decade rose from a fairly low to a medium level among the OECD-countries (Vuori and Vuorinen, 1993, 23-29). A simple but nevertheless very telling indicator of the same trend has been the changes in the occupational structure. A customary division for analyzing social development in terms of the occupational structure has been the three-sector model consisting of agriculture or the primary sector, the processing industry or the secondary sector, and the service industries or the tertiary sector. With the development of an information industry and technology a quarternary sector consisting of information occupations (Kennesey 1985) has been to become identified as a separate entity.

There are many other kinds of indicators for the development in the national production system than the model based on changes in the occupational structure. Industrial output, technology transfers, the number and kind of persons graduating from different educational levels, the number of patents, information about productivity etc. can all be used for describing some aspects of the technological development but as a shorthand description and an over-all measure the four-sector model will here be used as a main indicator of the technological development.

In the following a general description of the developments in the Finnish society will be presented in terms of the simple indicators here chosen.

2.2 The Decline of Political Mobilization

As stated, the rate of participation in the general parliamentary elections is here taken as the main indicator of political mobilization. Unlike the development in other Nordic countries the electoral participation in Finland has undergone a very uneven development. As the first European country with universal suffrage both for men and women, Finland had its first general parliamentary elections already in 1907. It was an occasion of great and intensive political mobilization, and the voting rate of 70.7 per cent was for those early times of universal suffrage exceptionally high. The voting turnout in these first elections was not reached and exceeded until the first elections after the Second World War in 1945, when voting participation rose to 74.9 per cent. In the 1920's the voting participation was below 60 per cent in all elections, and in the 1930's the highest rate of participation was 66.6 per cent in 1939. As a contrast the electoral participation in Denmark, Norway and Sweden showed an undramatic but steady increase all through the decades preceding the Second World War.

Also later there has been considerable fluctuations in the voting participation in the Finnish society. The main feature is a clear decline of voting participation in recent elections. In the two last parliamentary elections in 1987 and 1991, the voting participation was 72.1 per cent, the lowest voting turnouts after World War II.

In a parliamentary democracy with a multiparty-system a trend towards disinterest in politics is a highly problematic feature. It testifies to an increase in political alienation, which appears even more problematic considering that the youngest age groups eligible for voting abstain from voting more than others (Martikainen and Yrjönen 1991, 113).

2.3 The Development of the Web of Interest Organizations

During the first phase of industrialization there was a clear increase in the pattern of social differentiation. A separation took place between family ties and occupational roles. Individuals became less and less predestined to certain occupations and societal roles. Yet, it seems as if new hindrances to social differentiation have arisen. Political party membership has been the basis for many appointments of tenured civil servants even in fairly modest positions. A major development in the Finnish society in recent decades has been the rise of new interest organizations and a process of unionization. Traditionally Finland was known for a very low degree of membership in labor unions. The unionization rate in the 1930's was 10 per cent, the lowest figure ever recorded in a market economy. As late as in 1960 only a third of the labor force consisted of union members, and in the middle of the 1960's the unionization was still below 50 per cent of the total labor force. Then almost suddenly a great wave of unionization began. From the second half of the 1960's there was a steep increase in the union membership rates. Already in 1978 the big labor unions including both industrial and white collar workers covered more than 80 per cent of the total labor force. The current membership in the beginning of the 1990's is 87 per cent, one of the highest in the world (Kauppinen, 1992, 303).

Yet, unionization in itself is not a remarkable feature nor a source of grave problems. Industrial workers are organized all over the world, and the accomplishments of the labor unions through collective bargaining have often had clearly positive consequences for productivity. However, there are some additional features which are both remarkable and problematic. First, the strength of the unions as well as of other interest organizations has not only resulted in higher wages for their members but it has led to a rigid system of legal entitlements and rights which are hard to change with altering economic circumstances. Second, the great wave of unionization not only concerned the working class but there was also a rapid unionization in both the lower and upper middle class. Being organized in an interest organization became popular and the general rule even in groups such as professors, artists, and research council researchers in which individual incentives and competition are basic for their activities. Third, the Finnish farmers had been strongly organized already before the great wave of unionization started. As a consequence the farmers have been sheltered by fiscal arrangements which are hard to change even when the position of the agriculture in the national economy changes. Despite the high degree of unionization it almost goes without saying that the very weakest groups in the society such as the unemployed, the part-time workers, the foreign immigrants, and temporarily employed researchers are not organized. At any rate, the great wave of unionization and the rise of interest organizations of different kinds have led to a rigid system of entitlements constituting a hindrance to social differentiation.

2.4 The Rapid and Ruptuos Changes in the Occupational Structure

Almost up to the 1960's Finland was a predominantly agrarian country. It resembled the countries in Eastern Europe, and its path of development differed clearly from the other Nordic countries. In the mid-1960's Finland rapidly developed into an industrial nation. The most thorough change doubtlessly occurred in agriculture. In 1950 around 46 per cent of the Finnish labour force worked in agriculture but 25 years later, in 1975, the corresponding percentage was only 15.

The great structural change of the 1960's and 1970's was in the first hand observed as a great wave of migration. The Finns moved to cities, especially the Helsinki region, and to Sweden. Around 125 000 Finnish citizens moved during the second half of the 1960's to Sweden. The greatest immigration wave both in terms of numbers and consequences occurred in southern Finland. During the years 1951 - 75 the Helsinki region received a net gain of almost 300 000 persons from other parts of Finland.

In an international, comparative context it appears that rapid structural change in the 20th century has been an attribute of countries located in the peripheries (Alestalo, 1986, 34 - 39). During 1960 - 80 the decline of the labour force in agriculture and forestry was in Finland the second largest among all the OECD countries. The decrease of the labour force in the primary sector was largest in Greece, while the corresponding decline in Finland was roughly on par with what happened in Spain, Japan, Ireland and Italy. A clear decline although smaller than in Finland occurred also in Portugal and Turkey. The structural change of the Finnish society in the 1960's and 1970's is an illustration of a general regularity. Development and modernization in the peripheries often lag behind, but when they occur they come with force and high speed.

The structural changes in the occupational structure altered dramatically the class structure of the Finnish society. Undoubtedly, the greatest change was that the agrarian proletariat all of a sudden became almost non-existent. The issue of the social conditions in the countryside, the living conditions of the landless population, and later of the small farmers had been for hundred years by far the greatest social problem in Finland. It had given rise to conflicts, radicalism and agrarian reforms. Suddenly the agrarian proletariat disappeared before the very eyes of astonished politicians and economists (Alestalo, 1986, 63-71).

In 1950, over a quarter of a million people worked on farms with a cultivated area less than 5 hectars, but in 1985, this number had shrunk to 30 000, and in 1980 to only 16 000 individuals. At the same time the agrarian workers decreased in number, disappearing almost completely, while the number of persons engaged in forestry significantly diminished. Even today, small farmers and agricultural workers are poorer and less healthy than other groups in Finnish society. The rural proletariat has ceased to be a politically explosive and central issue but the group is still an important target group for social policy.

The great changes of the 1960's and 1970's had considerable human costs in terms of social and personal distress. Many categories of persons became socially uprooted. The change did not affect only those who moved to the densily populated urban areas in southern Finland or to Sweden. It also hit those who stayed behind in rural areas becoming almost devoid of human and social activity and those who lived in the target areas of migration and who had to adjust to the fact that newcomers began to form the majority in their local community.

Only afterwards has it become understood what the rapid changes in the 1960's and 1970's meant in terms of human costs. There was, for instance, a marked increase in the number of homeless "winos". In a comparative study carried out in the early 1970's on welfare and human relationships in all the Nordic countries, the number of completely isolated individuals with no social contacts whatsoever was significantly greater in Finland than in the other Nordic countries. According to this comparative study, 26 per cent of the adult Finnish population had not a single good friend outside the nuclear family. When this question was asked in a Gallup poll ten years later in 1982, the category of people with no friends at all had gone down to 12 per cent, which is the general level in the other Nordic countries. Other indicators show in a similar fashion the great human and social costs of the wave of the sudden change and migration (Allardt, 1985, 62-65).

The structural changes, however, did not stop at the great movement from the countryside and the rural occupations in the 1960's and 1970's. It is important to note that during the same decades the labour force in the secondary sector, the processing industry had already began to decline in the advanced countries of Western Europe. Countries such as Belgium, Great Britain, Holland, Sweden, Australia and Switzerland experienced during the period between 1960 and 1980 a considerable decrease of the labor force in industrial work. In Finland there was still at that time an increase of the labour force in industry but in the 1980's a downward trend of the number of people in the industrial sector began. In 1975 26 per cent of the labour force worked in industry but in 1985 it has declined to 23 per cent. An even more substantial change was the growth of the service sector. From 1980 to 1985 the industrial work force decreased by 30 000 individuals only but at the same time the work force in the tertiary or the service sector increased by 100 000 individuals. In Finland as in the other Nordic countries employment in the service sector is now greater than in industry (Allardt, 1988, 175-181).

Even more profound were the changes in the nature and quality of work. This can be observed when the four-sector model, introducing a quarternary or "information occupations" sector, is used. It has been shown that when applying Marc Porat's (1977) criteria for defining information occupations the quarternary sector had become larger than any of the primary, secondary and tertiary sectors already in 1985 (Paakkolanvaara, 1988, 94-96).

To sum up: in less than forty years Finland passed from an agrarian to an information society. It is obvious that such a dramatic change also invokes considerable human and social costs.

The changes in the occupational structure have here been taken as an indicator of rapid technological change in the Finnish society. Also other indicators, as seen from other presentations in this volume, indicate that the technological level in Finland has risen considerably in the last three decades. Finland adopted technologies developed elsewhere while at the same time increasing its own research efforts. Yet, the question arises to what extent the changes have been beneficial for the society and the economy. There have in Finland been deliberate efforts to raise the level of technology as part of a national strategy. The increase of the technological level has been considered as a goal in itself and as a source of national pride, but the question remains: how much can the technological level be raised without corresponding changes in other societal realms?

2.5 Problems of a Frozen Party System

Political scientists have already for several decades spoken about "the frozen party system". The background for using such a metaphor is the fact that while the social structure in the West European countries has undergone profound changes during the decades after the Second World War the party system and the strength of the dominating political parties have remained almost intact. Thereby it has become questionable whether people's aspirations and existing social conflicts are properly translated into the party system and political representation anymore.

The question of whether the party system is frozen is indeed very relevant when analyzing Finnish political life. The Finnish social structure underwent, as shown, profound changes during the last four decades and actually passed from an agrarian not only to an industrial but also to an information society. There have not been corresponding changes in the Finnish party system. Some changes, however, have occured. The Finnish Communist Party, which in the 1950's received something between a fourth and a fifth of the total vote, has more or less disappeared from the political scene.This is due to changes in both the international order and Finnish social structure. With the disapperance of the rural proletariat and the decrease of the labour force in certain industries the former Communist electorate has almost ceased to exist. Otherwise, however, the party system is to a considerable extent the same as it was several decades ago.

The low voting participation with 28 per cent of the electorate abstaining from voting is at least partly a sign of the fact that large groups of voters do not have political representation of their interests. In the local elections held in October 1992 the Greens made considerable gains in some of the most urban municipalities, such as Helsinki, while in some other municipalities different groups of independents succeeded in being elected. Nevertheless the non-voters were close to 30 per cent. As indicated, a most problematic feature is the fact that the non-voters are frequently young persons, in the age group from 18 to 24 (Martikainen and Yrjönen, 1991, 113-115). It is hard to avoid the conclusion that the present party system does not respond to the interests and needs of many of the voters. The question is to what extent the elections in the present situation really gives possibilities for ascertaining what people really want?

2.6 The Rigidity of the Web of Interest Organizations

Simultaneously as the ideologically based political parties have lost ground, as have also culturally oriented voluntary organizations, there has been an upsurge of economically and instrumentally oriented interest organizations. As mentioned above, the degree of unionization in Finland is among the highest in the world. In briefly discussing the implications some ideas from texts of the well-known economist Mancur Olson will be borrowed. In "The Logic of Collective Action" (1965) Olson points out that the services of associations such as labour unions, professional associations, cartels, farm organizations etc. resemble the basic services of the state in the sense that the *public goods* they provide go to everyone in the group in question. The members they serve have no rational incentives to contribute involuntarily to their support. Accordingly, organizations of this kind, especially when they represent large groups are not supported because of the collective goods they provide, but they are supported because of what Olson labels *selective incentives.* A selective incentive is one that applies to the individuals selectively depending on whether they do or do not contribute to the provision of the collective good. There are both *negative* and *positive selective incentives.* Negative incentives include, for instance, the coercion to pay dues and simply the inability to get a job unless one is a union member. Positive incentives in many organizations are exemplified by favourable insurance policies, publications, group air fares, and other things made available only to members. In a later volume "The Rise and Decline of Nations" (1982) Olson spells out some of the implications of the rise of interest organizations for the functioning of national economies.

Of special interest here are some of Olson's conclusions concerning the effects of different kinds of interest organizations. He makes a distinction between big *encompassing* and small *special-interest organizations.* The incentives facing an encompassing organization are dramatically different from those facing an organization representing only a narrow fragment of society. The members of a highly encompassing organization own so much of the society that they have an important incentive to be actively concerned about how productive it is. They are, in fact, in the same position as a partner in a firm that has only a few partners. Olson (1982, 48) gives a fictious example about an organization representing a third of the income-producing capacity in society. Such an organization will, on the average, "bear about a third of any loss in the society's output that results from the policies it obtains". And Olson continues "Thus any effort to obtain a larger share of the national income for the clients of such an encompassing organization should not make sense if it reduced the national income by an amount three or more times as great as the amount shifted to its members ...this can be a constraint of great practical importance. Clearly the encompassing organization, if it has rational leadership, will care about the excess burden arising from distributional policies favorable to its members and will out of sheer self-interest strive to make the excess burden as small as possible".

The situation is entirely different in small special-interest organizations. There is, in principle, no constraint on the social cost such an organization will find expedient to impose on the society in the course of obtaining a larger share of the social output for itself. It is hard to avoid saying that Olson's conclusions correspond to experiences from the Finnish society. Encompassing organizations such as the Central Organization of Finnish Trade Unions (SAK) and Finnish Employer's Confederation (STK) act with considerable constraint in presenting demands. In contrast unions and interest organizations representing special groups, such as airplane pilots, doctors, teachers, nurses, technicians, and crewmen on ice breakers, have little or no incentives to make any significant sacrifice in the interest of the society. In labour struggles they simply tend to argue about their rights to seize a larger share of the society's production. The same may be true for farmers and some groups of workers such as metallworkers, longshoremen, and truck drivers. The intention here is not to give evidence about which groups have acted without constraints in defending their own interests but to indicate that there in general seems to exist a clear difference between encompassing and special-interest organizations. The latter do not have to care about how they influence productivity or about what costs they invoke for having the system running.

In the Finnish society the attention should not be directed only towards unions and interest groups. It is to be noted that industrial companies and even whole production sectors may function in conditions resembling those described by Mancur Olson. The share of investments within the gross domestic product is in Finland one of the highest in the world (EVA, 1992, 19). It is also known that Finnish development and industrial policies have been very national, if not clearly nationalistic (Vuori and Vuorinen, 1993, 18-20). Even when such aims are not explicitly stated Finnish investment policies have been led at least partly by nationally developed images and idiosyncracies. It gives some industrial sectors special advantages of the kind bestowed upon special-interest organizations. This is a pattern which definitely needs further study but it seems correct to mention the agricultural sector as a kind of a special-interest sector. The investment ratio for agriculture has in Finland been exceptionally high (Pohjola, 1992, 42-47). This has been so despite the fact that not only has the population in the primary sector rapidly decreased but that there also has been a clear decline of the contribution of agriculture to the GDP (EVA, 1992, 17). Agriculture has in a sense developed from an encompassing sector to a special-interest sector. This is a situation in which the representatives of agriculture can drive their interests in terms of a combination of self-interest and national values without considering their contribution to over-all productivity. An open question is whether and where one can find similar cases among the different branches in the processing industry.

At any rate, the great surge of unionization and rise of interest organizations in the 1950's and 1960's meant if anything that entirely new groups became organized. Unionization became extended to entirely new sectors in the middle class and highly educated groups. Many of them represented fairly small groups, and they clearly developed into special-interest organizations. As a result the legal web of entitlements and rights grew considerably. This development was certainly also one of the causes behind the very rapid growth in the government expenditure in Finland. In the 1980's public spending in Finland grew more rapidly than in Sweden and in the EC countries taken as a group (EVA 1992, 35). Quite typical for the new Finnish labour market and many occupational categories is that many entitlements do not have any clear relationship to the work itself, that they extend the number of vacation days farther than in most countries, and that many rules invented constitute hinders for voluntary and creative accomplishments. Yet, in many fields of work, for instance in most fields of teaching and health care, new developments require that people do more than they have to do. In any case, there is a web of entitlements and rights which do not influence productivity in any positive way but which on the other hand considerably increase the costs of running businesses and the public sector.

Today in 1993 the government makes strong efforts in cutting the budgetary expenditures of the public sector. It is uncertain whether these cuts have the right address. They mostly amount to an elimination of jobs in the public sector instead of cleaning the vast system of both special entitlements in the public sector and nationally oriented but uneconomically directed subsidies in various industrial sectors.

2.7 The Indiscriminate Urge for Innovations

In the 1980's the Finnish economy overheated. New firms, activities and production lines, which later run into financial crises, were created. Of course, it is known that all resources should not be spent at once, and that in expanding companies, and production lines one has to secure a certain amount of economic solidity. Nevertheless there was in Finland in the 1980's almost a blind urge for expansion. How it came about and what were the driving forces is very unsatisfactorily researched. It seems, however, appropriate to ask whether the development at least partly was invoked by a deceptive and "unholy" alliance between pure market principles and traditional nationalistic development policies. The market principle gave the incitament to expand and the nationalistic development principles gave both legitimization and official support to everything indicating rapid development.

From individual psychology we know very well that too rapid change, or in other words an overwhelming amount of new stimuli, are detrimental for the individual. An excessive amount of new things or stimuli in the environment create either strong defence reactions, or they lead very rapidly to satiation with many kinds of negative effects. Some sort of *optimal stimulation* (Hebb, 1955, 243-254; Littunen, 1958-59, 152-153) is the socially and humanly most advantageous predicament. This is already conventional knowledge in psychology but it also fits very well with the wisdom exerted by good organizational administrators. They tend not to use all resources at once, and they do not use or expend at once all possibilities to expand their organizations. Also in whole societies one can observe how the pace of consummating stimuli can be too rapid.

There has in Finnish technology policy during the last few decades been a strong and almost blind emphasis on nationally supported innovations and activities promoting innovations. It is uncertain whether such a technology policy in the long run sustains economic development. One should ask whether the investment policies and routines have been too much directed by national idiosyncracies, and whether the technology policy of the other hand has been too much based on a conception that technological innovations of almost any kind are nationally valuable.

The word "innovation" has lately been used in several government and committee reports in a very general and unprecise way. A graver error than the loose application of the word itself has been the neglect to study the consequences of different kinds of innovations. When it comes to the nationally oriented technology policy mainly focussing on product development and innovations with short-time effects it seems both reasonable and almost mandatory to ask for a consideration of the social effects of the innovations supported. In Finland's present economic situation it would be of particular importance to consider the employment effects of the innovations supported.

There is a famous saying by Knut Wicksell presented in his lectures already in 1908 (1934, p. 164): "The capitalist saver is thus, fundamentally, the friend of labour, though the technical innovator is not infrequently its enemy. The great innovations by which industry has from time to time been revolutionized, at first reduced a number of workers to beggary, as experience shows, while causing the profits of the capitalists to soar".

And Wicksell continues by saying that certain types of innovations might have a negative effect on the marginal productivity of labour. Thereby they reduce the real wage. In such a situation the technical innovator works against the interests of labour.

Accordingly, there are technical innovations with very different kinds of consequences. From a technological point of view they may be equally good and stimulating but they may not be neutral from the point of view of economic and social policy. A neutral innovation from the economic and social policy point of view is one which increases the marginal productivity of both factors of production, capital and labour, in the same proportion. Labour-saving technical progress increases the marginal productivity of capital more than it increases the marginal productivity of labour, and capital-saving innovations increase the marginal productivity of labour more than they increase the marginal productivity of capital (Södersten 1980, 131-148).

The Finnish technology and innovation policy has not been able to prevent the Finnish economy during the first years of the 1990's from experiencing its worst recession and highest unemployment rates since the Second War. In the beginning of the year 1993 there is an unusual situation with both increasing exports and rising unemployment. Finnish technology policy might need a thorough reappraisal and overhaul. One of the questions reasonable to pose is whether government agencies supporting the work for technical innovations should direct their monetary support to capital-saving rather than to labour-saving innovations. At any rate, a serious question for consideration in Finnish technology policy is whether agencies supporting technological innovations should be much more clearly related to employment policy than has been the case.

References

Alestalo, M. (1986), Structural Change, Classes and the State. Finland in an Historical and Comparative Perspective. Helsinki: Research Group for Comparative Sociology, University of Helsinki.

Allardt, E. (1985), Finnish Society: Relationship Between Geopolitical Situation and the Development of Society. Helsinki: Research Group for Comparative Sociology, University of Helsinki.

Allardt, E. (1988), "Sosiaalinen rakennemuutos" , Academia Scientiarum Fennica, Yearbook 1987-88, 175-182. (In Finnish; English Summary, "Changes in the Social Structure").

Deutsch, K.W. (1961), "Social Mobilization and Political Development", American Political Science Review, Vol. 35, No. 3, 493-514.

Eisenstadt, S.N. (1966), Modernization: Protest and Change. Englewood Cliffs, New Jersey: Prentice Hall, Inc.

EVA (The Council of Economic Organizations in Finland) (1992), Indicators of the Finnish Society 1992/93. Helsinki: EVA.

Hebb, D.O. (1955), "Drives and the C.N.S.", Psychological Review, Vol. 62, No. 2, 243-254.

Kauppinen, T. (1992), Suomen työmarkkinamallin muutos. Helsinki: Työpoliittinen Yhdistys r.y., Julkaisu 1 (In Finnish; English Summary, The Transformation of Finnish Labour Relations in the Light of Corporatistic Choices).

Kennesey, Z. (1985), The Primary, Secondary, Tertiary and Quarternary Sectors of the Economy. Papers of the Nineteenth Conference of The International Association for Research in Income and Wealth, Nordwijkerhout, Netherlands.

Littunen, Y. (1958-59), "Income Security Values at Different Levels of Frustration". Turku/Åbo: Transactions of the Westermarck Society, Vol. 4, No. 4, 143-245.

Martikainen, T. and R. Yrjönen (1991), "Vaalit, puolueet ja yhteiskunnan muutos" (Elections, Policial Parties and Societal Change). Helsinki: Central Statistical Office of Finland, Studies 178.

Olson, M. (1965), The Logic of Collective Action. Cambridge, Mass: Harvard University Press.

Olson, M. (1982), The Rise and Decline of Nations. Economic Growth, Stagflation, and Social Rigidities. New Haven and London: Yale Universtity Press.

Paakkolanvaara, E. (1988), "Informaatioyhteiskunta ja informaatioammatit". Helsinki: Central Statistical Office of Finland, Studies 145 (in Finnish); English Summary, "Information Society and Information Occupations".

Pohjola, M. (1992), "Millaista Suomea olemme nyt tekemässä (What Kind of Finland are We Now Making)", Uudistuuko vai rapistuuko Suomi - hyvinvointiyhteiskunnan uudet valinnat (Is Finland Renewing itself or is it Deteriorating - the New Options of the Welfare State). A seminar report from August 27, 1992, Helsinki (in Finnish).

Porat, M. (1977), The Information Economy: Definition and Measurement, US Department of Commerce, Washington D.C.

Södersten, B. (1980), International economics, Second edition. London: MacMillan.
Wicksell, K. (1934), Lectures on political economy, Vol. 1. London: Routledge (Translated from the Swedish original, Föreläsningar i national ekonomi, 1908).
Vuori, S. and P. Vuorinen (1993), "Outlines of the Finnish Innovation System: The Institutional Setup and Performance", first chapter of this volume.

3 The Social Infrastructure of Innovation in Finland

Antti Kasvio
University of Tampere, Finland

3.1 From Technology-Oriented Studies Towards Comparative Analyses of National Systems of Innovation

One of the main limitations of traditional innovation research has been its narrow technological orientation. Researchers have been interested mainly in such things as the creation and utilization of new raw materials, technologies or products. A rather straightforward relationship has been assumed to exist between the growth of a nation's R&D expenditure, extensive technology programmes, effective diffusion of new technologies and the competitiveness of an economy (see e.g. Niitamo 1958, Rosenberg 1971 and OECD 1980).

Recently, the scope of innovation research has, however, expanded radically. Technological and organizational innovations have been seen to be closely interrelated. It is also recognized that an economy's innovative capabilities are in many ways connected with the entire social infrastructure within which the economy is functioning. Innovation researchers are increasingly interested in such things as the quality of a nation's scientific institutions, the level of its enterprise cultures, the performance of its educational system and in the capabilities of its labour relations system to support new development initiatives. This paradigmatic change crystallizes in the lively discussions under way at present around the 'national systems of innovations' (e.g. Dosi et al. 1990, Lundvall 1992, Nelson 1992).

Interesting results have been achieved of the extent and efficiency with which new technologies are nowadays applied in different advanced industrial countries. For instance, the IIASA 'Tes programme' reports (see Ollus et al. 1990), Boyer's comparisons between the USA, France, Japan, Germany and Sweden (1989), and Jaikumar's analysis of the uses of FMS in the U.S. and Japanese manufacturing industry (1986) can be mentioned here. All these studies seem to support the thesis that certain 'societal effects' exert a very strong influence on the manners in which new 'post-Fordist' manufacturing strategies are being utilized. It has also been recognized that these manners are deeply rooted in the entire cultural heritages of these countries (d'Iribarne 1989).

So far most of these discussions have been confined to comparisons between a few advanced industrial countries. Among these, the U.S. manufacturing industry has often been sharply criticized for the prevailing dominance of old-fashioned Fordist mass production thinking (see e.g. Dertouzos et al. 1989, Adler 1989). The United States has also been said to be lacking a well functioning vocational training system and a developed corporatist system of interest representation. As a national system of innovation the United States has, however, also certain important advantages on its side. These include, for instance, the very high quality and performance of its leading academic and scientific institutions and as the fluently functioning relations between business and academic research (see e.g. Nye 1990).

Japan has often been described as the foremost example of a well functioning national system of innovation. Even though Japanese research institutions have not always been the leading ones in basic research and large-scale science-based innovations, the national technology programmes effectively organized under the custody of the MITI have made Japan by far the world's most successful utilizer of new knowledge. Japan has also been praised for its strong enterprise cultures, the practice of life-long employment and high employee commitment. These features have been further strengthened by the system of company-level trade union organizations. Important factors are also the great attention paid to the continuous development of employees' skills and on the well functioning horizontal and vertical information structures within enterprises. It is important to notice that Japanese factories do not have at all 'unskilled' workers in the sense as this category is known in Western countries. They have only 'skilled' and 'not yet skilled' workers (see e.g. Dore 1987, Cole 1989, Lillrank & Kano 1989).

Germany has pioneered in organizing large-scale natural science research in huge industrial laboratories already since the end of the 19th century. This initiative has supported the strong development of Germany's chemical, machine-building and other key industries during the following decades (Freeman 1992). As a utilizer of new technologies German manufacturing industry enjoys nowadays the special advantages offered by its 'Facharbeiter' institution - both as a vocational training system and as an important element in the social hierarchies within enterprises. One has to take into account also the highly institutionalized corporatist labour relations, functioning via the specific 'dual' system of interest representation (see e.g. Streeck 1992, 137-168). Partly this description is also valid for the former East German area, where great stress was always laid on the 'polytechnical' training of industrial workers.

The Nordic countries have often been counted among those economies which have been able to offer a comparatively favourable environment for effective utilization of new technologies. With their neo-Keynesian economic policies and flexible adaptation strategies they have been able to stand fairly well the turbulences of world economic developments since the mid-70s (Katzenstein 1985, Standing 1988). Even though these countries' R&D potential cannot be compared with that of the large industrial powers, in the area of civilian research the differences are smaller. Especially Sweden has also been known for its high-level engineering culture. The Nordic countries have an extensive network of vocational training institutions, although the organizational forms vary fairly much within the region. Sweden and Norway have had a strong tradition of 'managerial humanism', which has made possible the famous sociotechnical experiments in Swedish and Norwegian manufacturing industries from the 1960s onwards (e.g. Sorensen 1980, Elam & Börjeson 1991).

In all Nordic countries the industrial relations are nowadays built on an extensive system of corporatist interest regulation, which has effectively reduced differences between the various groups of wage-earnes (see Pekkarinen et al. 1992, Bruun et al. 1992), and the reformist governments have launched several large-scale work reform programmes in order to improve the quality of working life in these countries (den Hertog & Schröder 1989, 17-32). This institutional environment has supported broad employee participation and collaborative approaches in the design and implementation of user-oriented technical and organizational solutions both within industry and in different post-industrial spheres of work (e.g. European Foundation 1989, Freeman 1989; for more general information on national systems of innovation in Scandinavia, see Edquist & Lundvall 1989).

3.2 Finland as a Special Case Among the Nordic Countries

Finland must be regarded as a slightly special case among the Nordic countries. The process of industrial modernization started in Finland somewhat later than in Western Europe, and since the Second World War, Finland has had a rather specific foreign policy and trade policy position, characterized by its pact of friendship and cooperation and extensive economic contacts with the Soviet Union and other Comecon countries. Finland's opening to West European markets and to broader world market competition has proceeded gradually (Finland became a full member of the IBRD and the IMF in 1948, the GATT in 1949, the OECD in 1965 and the EFTA in 1986), and Finland's macroeconomic policies were not very Keynesian in the post-war decades (Pekkarinen 1988). Neither have the institutions of an advanced welfare state or the corporatist industrial relations developed in the same rhythm in Finland as among her Nordic neighbours (Crouch 1990). As a matter of fact, Finland started becoming a modern welfare state, comparable to other Nordic countries, only since the late 1960s, and the impact of social democracy on the development of her institutions has not been as dominant as in Scandinavia (on the development of a welfare state in Finland,see e.g. Alestalo & Uusitalo 1986).

Even nowadays Finnish industry depends very much on its forest sector. The paper and pulp industry and the manufacture of forest industry machinery are the leading branches of Finnish export industries. Although the productivity in Finnish industry and its research intensity have risen fast, especially during the 1980s,

Finnish industry is still lagging behind its leading Western competitors (Rehnström 1992). Only recently the share of high-technology products in Finland's industrial exports has started to gain some significance, and the electrotechnical industry has become Finland's most expansive export branch. An important role in this process has been played by the large-scale national technology programmes launched by the Tekes, the development agency within the Finnish Ministry of Trade and Industry (Vartia and Ylä-Anttila 1992, 100-102; Landesmann 1992, 261).

In general, Finland's academic research and education can nowadays be said to be fairly well- equipped, even though especially during the 1980s, there was a certain conflict of priorities between the funding of basic research - i.e. universities and the Academy of Finland - and the more applied research and development activities via the Tekes and the Ministry of Trade and Industry (Allardt 1990, 629). Finland has also a very extensive network of universities, which is dispersed over the whole country. Its problems are, however, the small size and relative isolation from the international scientific community. The technical universities and medical faculties have fairly well developed industrial contacts, whereas in more humanistic disciplines the cultural distances between business and academic research are still rather great. At present, there is a lively debate under way in Finnish university politics about the potential advantages of reducing the number of higher educational institutions and about the needs to improve their performance and internationalize their activities (see e.g. Numminen 1992).

General responsibility for the development of science and technology policies in Finland is borne by the Science and Technology Council, a high-ranking body chaired by the Prime Minister. Traditionally, the focus of its policy statements has been on the national R&D system. In recent years, however, the OECD report on new technologies in the 90s (OECD 1988; the group of experts that wrote the report was led by Mr Ulf Sundqvist from Finland) and its follow-up conference held in Helsinki in 1989 (Freeman 1989) have exerted much influence. Thus, in 1991 the Science and Technology Council issued a new review in which it adopted as its starting point a much broader view of the 'national innovation system'. The latter is defined here as the totality of all factors influencing the development of new knowledge and skills - science and the educational system being the most important ones - and which are able to function as a kind of a toolbox for the production of social, human and technological innovations (Valtion... 1991, Seppälä 1991).

Finnish enterprise cultures have originally emerged together with the industrial revolution, under the influence of such foreigners as, e.g., Thomas Finlayson and Hans Gutzeit. This culture has not, however, remained very international. Only during the 1980s the major Finnish export corporations started actively to internationalize their operations. There are some astonishing success stories - like the development of the Nokia corporation from a local producer of cables and rubber shoes into Europe's third largest consumer electronics manufacturer - (Ylä-Anttila & Lovio 1991), but, in general, Finnish enterprises suffer from their relative isolation from Europe's leading business centres and from relatively scarce international experience. Serious mistakes were also made when Finnish industrial corporations started to expand their international operations, and even in Finland the best results are at present achieved by producers functioning under foreign ownership. The layer of innovative small enterprises continues to be fairly thin in Finland - even though their emergence has been actively supported, e.g., by the creation of 'Science Parks' around all technical universities (see Vuorinen 1991).

Finland's network of vocational schools is at present one of the most extensive ones in the world, and nowadays Finland allocates a larger share (17% in 1988) of her public spending to education than any other OECD country (see Economist 1992, 16). The problem is, however, that a high level of expenditure does not necessarily guarantee the system's superior performance. In fact, one special feature of the Finnish vocational training is its strongly school-centred character. Thus it is not as well connected with practice as, e.g., the German system, in which essential parts of training are given within enterprises (Asplund 1991). Neither is the social status of skilled workers in factories as firmly established in Finland as, e.g., in Germany; as a matter of fact, recent studies have shown that Finnish employers often prefer workers without vocational training in their recruitment policies (see Silvennoinen & Pirilä 1992). The resources spent on labour-market and on-the-job training in Finland have also been fairly modest (Standing 1990, 205).

Nowadays there are some extensive reform programmes under way on in the Finnish vocational training system. One of them is the 'youth school' experiment. This means that the traditionally very clear line of demarcation between vocational training and the more general education given in schools of higher ecucation is to be radically reduced, and the students are given a possibility to take courses from both institutions. Another important new phenomenon are the 'polytechnics' or 'Fachhochschulen'. These institutions complement the tertiary education given in universities, and individual courses can be utilized vice versa, but the goals of these institutions are not scientific but vocational. New measures are taken also to develop adult vocational training and the training of personnel within the enterprises (Kämäräinen 1991).

In practice, Finnish enterprises are, however, still today far from developing into real 'learning organizations' in the sense given to this term, e.g., by Charles Handy (Handy 1989, 168-187), i.e. as organizations, in which the continuous development of employee's skills has become an essential part of these organizations' everyday life at all hierarchical levels. Recently, Finnish enterprises have, however - at least at a rhetorical level - been laying more stress on the development of their personnel and on new forms of cooperation between business and the educational system (see e.g. Kairamo 1989).

3.3 Industrial Relations in Finland

The Finnish industrial relations system has usually been characterized as representing the Nordic type of 'social' or 'democratic' corporatism. In several recent studies this type of industrial relations is supposed to have contributed positively to the good economic performance of these countries (see e.g. Katzenstein 1986, Pohjola et al. 1992 and Standing 1990). Also in this respect, Finland's development has, however, its special features. Finland had, as a matter of fact, a rather decentralized system of industrial relations up to the late 1960s, when the Finnish trade union movement was unified around the SAK and the general incomes policies were adopted. Since then, the unionization rate has reached a very high level - about 80% -, but the numbers of strikes and the days lost in them continued to increase considerably in the 1970s (see Kauppinen 1989, 98-104, Kauppinen 1992, 219-223).

The relatively late development of the broad corporatist interest representation system in Finland can probably best be explained by the particularities of her political history. Unlike other Nordic countries, Finland has had a strong Communist party with influential positions within the trade union movement. There existed still in the late 1960s and early 1970s a militant left opposition against the introduction of centralized incomes policies. Despite this opposition, highly concentrated corporatist arrangements were developed in Finland at that time, and even social legislation was uniquely connected with them via the 'social packages', annexed to collective wage settlements (Saari 1992, Piirainen 1992). Since the latter half of the 1970s, however, the Finnish industrial relations system has become increasingly tainted with the general atmosphere of 'consensus'. This means that there exists a strong tendency to solve potential interest conflicts both at the national as well as at the enterprise level rather via negotiations than an open use of force. Thus, for instance, fierce industrial conflicts have not broken out any longer in the traditional male-dominated 'workers' fortresses' (an interesting case is described in Niemelä & Leimu 1991), whereas several of these have been experienced within different female-dominated service and white-collar trades.

The character of industrial relations and managerial cultures are usually closely linked to each other. In Finland the prevailing managerial cultures have traditionally been much more authoritarian than, e.g., in Norway or Sweden. The Finnish industrial relations climate has perhaps been interpreted to require somewhat harsher methods of command than has been considered necessary in 'softer' Scandinavian environments. It must also be taken into account that several managers in postwar Finnish enterprises were recruited from the armed forces, and even the rest of Finnish managers had predominantly an engineering background. Thus, there was no basis for the adoption of more broadly 'humanistic' visions of managing people. Neither did the actual labour market situation create any vital needs for an adoption of new approaches, because - unlike in Sweden - the extensive internal migration secured a fairly good supply of labour force for Finnish industrial enterprises up to the 1980s (on the differences between the Finnish and Swedish managerial cultures, see, e.g., Laine-Sveiby 1991).

These kinds of factors probably also explain at least partly the fact that the international 'socio-technical' movement did not yet gain any significant influence in Finland in the 1970s, although Finland's neighbouring countries, Norway and Sweden, were among the pioneers in adopting new organizational principles. The ideas of 'humanizing' work were suspected both by Finnish enterprise managers, strongly oriented towards the traditional thinking of rationalization, and by the trade union militants, who often regarded humanization simply as another way of exploiting and manipulating the workers. Therefore, at the same time as the new forms of work organization were broadly experimented in Swedish and Norwegian factories, and widescale national programmes were launched in Western Europe to support these experiments, Finnish progress was very modest (see Hänninen & Kauppinen-Toropainen 1981), and the main focus of reform activities was laid on the development of labour protection, occupational safety and other 'concrete, material issues' (see Kasvio 1991). Within the Finnish academic social science, humanization was also discussed strictly as a theoretical issue (see e.g. Julkunen 1987).

The traditions of corporatist regulation of employment and of broad employee participation have not been as well rooted in Finland as, e.g., in Sweden or Germany, either. As a matter of fact, it could be said that among West European countries Finland has perhaps been one of those in which enterprises have had the best opportunities to adapt their personnel into changing market situations (Emerson 1988; for a Nordic comparison, see Nielsen 1992). Neither have the participation rights been very extensive, even though the Act on cooperation within enterprises was issued after lengthy preparations in 1979. This law did not, as a matter of fact, establish any co-determination rights for employees. Rather, it obliges enterprises to inform their personnel beforehand about certain decisions, if these are considered to have important consequences for their work and employment situation. Thus, the system of elected shop stewards and labour protection delegates and of rationalisation cooperation have had, in practice, much greater significance as participatory channels than the bodies formed on the basis of the Act on cooperation (e.g. Flodgren 1992).

Recently, some progress has been made both at the legislative level and the level of actual managerial practices. The 'Finnish work reform' was started by the Government in 1987 with the target of bringing Finnish labour institutions to the European level. Subsequently, Parliament has passed new legislation on, e.g., 'Employee Investment Funds' and on employee representation in the decision-making bodies of enterprises (see Kasvio 1990, Lilja et al. 1990, 219-220). A new generation of progressively thinking managers has entered the stage in enterprises, and the management training institutions have campaigned actively in order to change leadership cultures into a more democratic direction (Allardt 1988). Also the sociotechnical organizational thinking has begun to gain new ground in Finland in the late 1980s and early 1990s, and at present, there are active preparations in progress for the launching of a new national QWL programme (see Komiteanmietintö 1991). A good example of enterprise-level changes is the Finnish clothing industry's quick move from technologically oriented strategies in the 1980s towards an active search for organizational flexibility via an extensive utilization of teamwork in the 1990s (see Lavikka 1992).

These changes in the Finnish labour policy regime have partly been connected with expectations according to which an increasing scarcity of labour force would be becoming a major problem in Finland in the 1990s. Due to the present recession this prognosis will hardly be realized; it can rather be expected that unemployment will remain high for several years to come. However, the new climate in which new work reforms are discussed has survived at least for the time being.

3.4 Policies of New Technology in Finnish Enterprises

I have mentioned earlier that a collaborative approach to the design and implementation of new technologies and procedures has often been considered as being an important feature of a well-functioning national innovation system. At present there are different interpretations concerning the policies of new technology in Finland. It can be stated, e.g., that within the mainstream of managerial tradition major technical changes have been decided mainly in the upper echelons of the enterprises, whereas employees have been informed about the decisions taken at a relatively late stage. Nor does there exist any generalized policy of negotiation between employers and unions over changes in work organization (Lilja et al. 1990, 206).

On the other hand, there are some famous examples of the adoption of more participating approaches to technological changes in Finland. There is, for instance, the paragraph 11 in the collective agreement of the Finnish paper industry. According to this paragraph, employers are formally obliged to start negotiations with trade union representatives whenever new arrangements are introduced which can be thought to influence the working conditions at some stages of the production process (Alasoini 1992, 360-361). Fairly much has also been written about the positive experiences of controlled change in the Finnish graphical industry where a special technological agreement has fixed the employment procedures which are followed in the change processes. A third well-known example has been the way in which new technologies have been introduced in the Finnish banking sector (see e.g. Murray 1989).

In addition, for instance, during the late 70s and early 80s, when there was a lively debate about the opportunities and risks connected with the introduction of new technologies, Finnish trade unions took generally an exceptionally positive stand towards techical development. This was reflected, e.g., in the discussions of seminars organized by e.g. the SAK and the Finnish Metal Workers' Union on new technologies (see Vuori & Vuorinen 1992, 18-19).

There are, however, also more critical interpretations of the policies of new technologies in Finnish enterprises. For instance, Pertti Koistinen and Kari Lilja are unwilling to talk about a 'collaborative' approach in the sense that the partners would be joining their interests on an equal basis. Instead, they see the 'consensual adaptation' of employees and their representatives to a change in the working conditions as being the dominant approach in today's Finland (Koistinen & Lilja 1988; see also Alasoini 1991).

Kimmo Kevätsalo, a research specialist of the Finnish Metal Workers' Union, has, on his part, analyzed the policies of new technology within the shopfloor trade union organizations of large Finnish metal industry corporations. Kevätsalo has identified plenty of suspicions and defensive attitudes towards the new organizational solutions proposed by management, even in those cases in which these solutions have opened also significant new opportunities for improving the position of workers in these enterprises. Kevätsalo's main interpretation about the situation is that the local cultures of interest representation have traditionally focused their attention almost solely to the 'distribution issues', and that is why they are not without long and difficult learning processes very capable of handling successfully those 'production issues' that are gaining increasing importance within factories (Kevätsalo 1991).

Matti Kortteinen has found in his recent dissertation that the new managerial strategies adopted in Finnish industrial enterprises do not always produce exactly the expected results. On the contrary, the skilled workers who have started enthusiastically their work with programmable machine tools and FMS stations find often themselves deeply disappointed with the actual realities of their work situation. Very often they react also by returning to the old strategies of getting as high piecework rates with as little effort as possible. This happens especially when the organizations stick to the traditional social hierarchies which effectively prevent workers from ascending to the status they feel to be entitled to according to the character work they are doing (see Kortteinen 1992). Even if Kortteinen has been criticized, e.g., for having interpreted too narrowly the significance which the learning of new skills has for the workers (Toikka 1992), his analysis reveals undoubtedly important potential sources of conflict in modern industry.

3.5 The Innovative Potential of Finnish Cultures of Work

Highly developed and innovative cultures of work can also be counted among the necessary preconditions for any nation's competitiveness. The Finnish economy is able to enjoy the advantages of a strongly established culture of work. This can be seen, e.g., in the high active female labour force participation rate (with a large majority of women being in full-time employment). Also according to recent comparative surveys, work is a central life interest for a relatively large share of Finns - both women and men -, although we have been moving closer to the more typical European patterns in this respect. Finns also lay much stress on accomplishing something important in their work (e.g. Lotti 1985). But although Finland has sometimes been called the most Lutheran country in the world - which, therefore, ought to be highly committed also to the values of protestant work ethic - , the regular working hours of the Finnish industrial workers are among the shortest in the world (1.723 hours in 1992), and in the 1980s certain pathological behavioural features have come to the fore. For instance, men in Finland retire, on average, earlier (at the age of 58 years) in Finland than in most other European countries (Poikkeus 1992). There have also been plenty of discussions about the young people's 'cool' attitude towards work (Tuohinen 1990), and the national QWL-surveys have shown an increasing dissatisfaction with their jobs among Finns (Kolu 1991, Lehto 1991).

Perhaps the most interesting interpretation of the peculiarities of today's Finnish culture of work has, however, been presented by Kortteinen in his dissertation. According to Kortteinen, it is typical of Finns to think that 1) work is hard, that 2) one has to fight one's way through the difficulties faced at work, and that 3) one has to accomplish this with one's own resources because, basically, it is a question about one's honour (Kortteinen 1992, 43). This culture is not necessarily a particularly happy one, nor is it always very creative or team-oriented. Nevertheless, the existence of this culture with its behavioural consequences must be taken into account when new managerial strategies are being developed.

3.6 Towards a New Era of Industrial Development

In the 1990s, advanced industrial economies are entering a new era in their development. New flexible modes of production are paving their way through. This means that also new requirements are set on the production personnel, and their skills are used much more broadly than before (e.g. Thurman 1991). Some authors have interpreted this to lead to the formation of entirely new kinds of horizontal and vertical information structures, closer interconnections between the design, manufacturing and distribution departments and to changing forms of enterprise-level industrial relations. For instance, Aoki has postulated that in modern industrial enterprises the higher wages paid to skilled personnel can no longer be understood as purely labour costs in the traditional sense of the word. It is rather a dividend paid for the collective skill capital of the personnel. Also the management's role is changing: it is no longer functioning as the owner's representative with respect to personnel, but rather as a mediator between the interests of the owners of its physical and knowledge capital (Aoki 1986).

There are different and partly competing models of flexible manufacturing. One of these is the model of 'lean production', described by Womack, Roos and Jones in their recent publication (Womack et al. 1990). This model is very much built on the experiences of the Japanese automobile industry. In Western Europe there has, however, emerged also a lively debate around the 'anthropocentric' or 'skill-based' production, whose main ideas come from advanced machine-building industry (see e.g. Lehner 1991). Although these two models have very much in common - e.g., a great stress laid on teamwork, close interaction between design and manufacturing etc. -, there are also many significant differences between the two approaches. The 'lean production' concept is, for instance, more oriented to leadership, whereas the APS lays more stress on participation; 'lean production' is organized mainly around existing technologies, whereas the APS wants to adapt also the technologies to the new requirements etc. The proponents of the APS often present also the opinion that their own concept would fit in better with the European corporatist industrial relations environment than with the concept of 'lean production'.

For the time being, Finland is not a member of the European Community, and thus she cannot officially participate in the research projects on anthropocentric manufacturing organized, e.g., under the auspices of the Monitor/FAST programme. This also means that there are no comparison data on the actual spread of APS principles in the Finnish manufacturing industry (for a European overview, see Wobbe 1992). Individual Finnish researchers have, however, participated in the initiatives that have contributed to the development of new EC research programmes on the topic (Ranta 1989), and at present, there are research programmes under way in which the potentials of new paradigms are being tested in practice in four Finnish manufacturing enterprises (Hyötyläinen et al. 1991). More generally, it can also be stated that the special structure of Finnish industry (which has never been strongly oriented to the Fordist-type mass production) is apt to experiment with the new strategies - which do not differ only from the traditional mass production model but also from the Scandinavian 'sociotechnical' model (see e.g. Kiviniitty & Alasoini 1992).

3.7 Finland at a Turning Point

At present, the Finnish manufacturing industry is in very special conditions of change. This is because after a period of vigorous progress in the 1980s the Finnish economy entered the most severe crisis of her history in 1990. In two years, Finland's GDP has receded by about 10%, the rate of unemployment has risen to more than 18%, property values have fallen dramatically, and the Finnish banking sector is experiencing a very severe crisis. All the major export corporations have made great losses, and their debt expenses are growing rapidly. After the depreciation of the Finnish markka by about one-third and two subsequent zero-level collective agreements, the competitiveness of Finnish industry has, however, improved considerably, and, at present, her output and exports are expanding rapidly (see e.g. ETLA 1993).

The basic issue is, however, that Finland is not undergoing at present only a normal cyclical crisis. Rather, the entire position of Finland in the postwar European system is changing. For several decades, Finland was able to develop a fairly sheltered enclave economy with good export prospects to the Soviet Union, large supplies of energy and raw materials plus unlimited opportunities to operate in Western markets. With the economic decline of the Soviet Union and other Eastern European countries - which is nowadays also complemented with increasing economic problems in the Nordic welfare states - Finland has to redefine her position in tomorrow's Europe. Especially high requirements are set on the performance of Finnish industry, because it has to retain its competitiveness in a situation in which European industry, in general, is in great difficulties and in which essentially cheaper facilities are becoming available in its nearby surroundings.

3.8 The Need for Radical Innovations

Finnish manufacturing enterprises have invested heavily in new machinery in the 1980s - with the expectation that this would raise their productivity to the level of their West European competitors. Because of their heavy debt burden and of the crisis of the banking sector, the Finnish manufacturing industry is not at present able to start new investments even if its sales were expanding. This means that better performance and increased flexibility must be reached mainly through organizational means. In this situation, the openness of the existing labour relations system to innovative solutions becomes vitally important.

On the macro level, we have to ask how the institutional setting of Nordic welfare states and corporatist industrial relations will adapt itself to the changing conditions. It has been originally developed to serve the requirements of Fordist mass production and of a relatively homogenous wage-earner society. If we were really moving towards a Post-Fordist, flexible network economy in which new 'atypical' forms of employment are increasing, can the old social infrastructure of production survive and be able to contribute positively to the dynamic development of the new economy that is emerging? Or must the Nordic welfare states be built down, will the climate of concensus and negotiated reforms degrade into deepening conflicts between the social partners, and will the implicit 'gender contract', on which the Nordic institutions are resting (e.g. Julkunen & Rantalaiho 1991) break down?

Several authors consider that in today's industrial revolution radical changes are required also within the social infrastructures of production (e.g. Perez 1983, Roobeek 1987). For the Nordic countries, the prognosis is that we shall be moving from the present forms of 'democratic corporatism' into the predominance of 'enterprise-level corporatism' or 'microcorporatism', in which wage negotiations are decentralized and the most essential decisions are taken within enterprises (see e.g. Brulin 1989, Hohn 1991, Kauppinen 1992, 184-209). This would naturally mean that also the differences between wage-earners employed in the more and less competitive part of the economy would be expanding considerably. It can be expected that the present large apparatuses of public welfare service provision in the Nordic countries shall be replaced by a system in which the state (national or local) guarantees certain rights to the citizens, collects taxes and uses them to order a certain amount of services outside the professional service organizations - either from public or private, but anyway from organizations which are forced to continuously improve their performance in order to maintain their share of the market (see e.g. Kosonen 1991, Normann 1992, Riihinen 1992).

One important question arising necessarily concerns the role of national policies in the development of tomorrow's economies and their innovative capabilities. As, for instance, Robert Reich has stated, traditionally the focus has been on improved competitiveness of the nation's leading enterprises. In the 21st century we are, however, moving towards an economy in which corporations will have no nationality. Capital will be increasingly mobile and searching for well-functioning environments for its different operations all over the world, and also raw materials, technologies and labour force will be moving rather freely across the national borders. In such conditions, a nation's competitive capacities will no longer depend on this nation's major enterprises (because there are not such), but rather on the social infrastructures of production it has been able to develop. A crucial factor will be the competitiveness of a nation's institutions of work, because the wealth of nations will increasingly depend on the efficiency with which it will be able to mobilize its labour resources (Reich 1991).

This logic has recently enlivened a new form of Keynesian thinking. The problem with traditional Keynesianism was that it saw the main task of public intervention in the regulation of total demand. It was a far too easy target for monetarist criticisms at a time when expansive policies led to soaring public debts, rising foreign trade deficits and the inflation of national currencies. At present, a new form of supply-side Keynesianism is gaining ground which sees the public sector's main tasks to be on the supply side, i.e. in the continuous improvement of the social infrastructures of production (Streeck 1992). This is a discussion which is becoming increasingly relevant in the current Finnish conditions in which new means must be found in order to solve the quickly aggravating problems of unemployment and new measures to promote equality within the increasingly deregulated labour markets will also be needed.

3.9 Finland as Part of the Baltic Sea Region

At present, it is timely to also ask to what extent the Finnish economy's problems are really national by nature, and to what extent they can be solved by developing the social infrastructure of production within a nation-state's framework. This question has been brought to the fore by the observations according to which the processes of dynamic economic development and stagnation are increasingly regional phenomena (Scott & Storper 1987). Thus, for instance, when examining the distribution of major scientific and technical innovations within the EC area, distinct regional differences can be identified (see e.g. Hingel 1992).

Looked from this angle, it is interesting to analyse Finland's position in the Baltic Sea region. At present, it is an area with many severe problems (an economic downswing in the post-socialist St. Petersburg area, Baltic republics and Poland; slow growth, unstable currencies and increasing unemployment in the Nordic countries; stagnation in the traditional industrial centres of Northern Germany; controversies between ethnic and national groups, military tensions, hatred and violent riots against foreigners). But in the future, this region has all the prospects of becoming one of the most dynamically developing parts of Europe. It will come closer to the heartlands of Europe when the centre of gravity of the European Union is moving eastwards (which I believe to happen sooner or later). It consists of areas with highly developed infrastructures (Nordic countries and Northern Germany), areas with abundant supplies of cheap and highly skilled labour force (the post-socialist countries), and it is connected (via St. Petersburg) with the world's richest sources of natural resources.

In order to realize the positive visions for this region, account should be taken of such issues as how to hinder the presently threatening economic division of Europe, to bring together the different competitive advantages potentially available, to maintain peace and democracy in the area, to develop living contacts between the region's metropolitan centres, to create well-functioning labour markets and institutions of work within the entire Baltic Sea region, and how to save the region from the huge ecological risks threatening its future at present. If positive solutions can be found to these problems, one could perhaps in the future start not talking about the 'national system of innovations' but rather about a dynamically developing regional system of innovations ranging from St. Petersburg up to Northern Germany. But in this connection it has to be taken into account that this kind of development does not automatically contribute positively to developments in all parts of Finland. Rather is it probable that the 'regionalization' of economic dynamism will increase the risks of unequal developments within the country, unless effective counteractive measures are taken.

3.10 Summary: Will the Nordic Democracies Retain their Competitive Advantages?

The analysis shows that at the end of the 20th century, Finland - together with the other Nordic countries - will enter a new world, radically different from the one that existed during the postwar decades of exceptionally high economic growth and nearly full employment. At that time, the Nordic welfare states with their corporatist regimes were able to act even globally as models of a 'good society', in which economic dynamism and highly developed enterprise cultures were in a unique manner combined with social solidarity and the respect of egalitarian values.

In today's transformation period it is reasonable to ask whether the basic features of Nordic systems of social corporatism are able to survive in the internationalized economy of the 21st century, and, if this is the case, whether they still have the capacity of contributing positively to the economic, social and cultural performance of these societies.

In recent discussions it has often been pointed out that, for instance, the ongoing West European integration processes do not necessarily lead to a convergence of the existing industrial relations systems and that also the Nordic social corporatism may therefore at least, in principle, be able to maintain some of its present specific features in the years to come (McDaid 1991, Bruun & Nielsen 1992, Streeck & Schmitter 1992). Thus the most severe challenges for the existence of Nordic corporatism will not be presented by the external environment but rather by internal sources, i.e. the break-through of new managerial strategies and the strong trends towards decentralisation and enterprise-level solutions in collective bargaining. But as, for instance, Pohjola has shown, the social relations of production and the collective skills emerging in tomorrow's enterprises will open entirely new outlets for collective interest representation and renewed forms of 'democratic corporatism' (see Pohjola 1988, Pohjola 1992, 77-78).

A much more contested issue is, however, whether the Nordic types of 'social' or 'democratic' corporatism will also be able to offer a fertile soil for a successful implementation of innovations in the emerging post-Fordist network economies. Many critics are nowadays in favour of a more open play of the market forces, the 'deregulation' of the functioning of Nordic labour markets and of an immediate elimination of the rigidities hindering the flexible adaptation of Nordic economies to the requirements of tightening international competition.

But there exist also different kinds of interpretations of this issue. According to these, deregulation is not necessarily the best way for the Nordic countries to adapt their institutions of work to the requirements of tomorrow's world economy. One has to take into account that several other economies would in any way be able to outdo Finland and other Nordic countries with lower labour costs and with much more deregulated labour markets. The specific competitive advantages of the Nordic institutions of work will not be found in less regulation but rather in their high level of progress, which can be reached only through a well- balanced mix of flexibility and regulation and through active labour political measures to guarantee a full utilization of these nations' human - female as well as male - resources.

As, for instance, Wolfgang Streeck has very convincingly shown in connection with vocational training, a fully deregulated system, in which effective market demand is the sole factor determining the supply of skills, would necessarily lead into continuous market failures. Moreover, in an era of flexible production these failures would contradict even the immediate interests of enterprises acting as subjects within the market. Thus, in order to produce an adequate supply of skills, a certain level of neo-corporatist 'regulation', i.e. an establishment of norms obliging individual enterprises to recruit a certain number of trainees - more than they would do acting solely on the basis of their own self-interest -, will also be necessary in tomorrow's conditions (see Streeck 1989).

This observation could quite well be generalized further, to describe the potentialities for a renewed Nordic corporatism in supporting the overall institutional restructuring of the Nordic societies - in close cooperation with other parts of the emerging Baltic Sea region - into well functioning post-fordist network economies. This goal cannot be achieved without the creation of social infrastructures that are actively supporting the adoption of innovative practices in all spheres of life, not only within the productive core of the economy. Perhaps through these processes of economic and social transformation the Nordic societies might sometimes in the future find themselves again in a position in which they are serving as models of highly performing but, at the same time, egalitarian and morally just societies for the rest of the world. But for this positive vision to be realized, a new consensus should be found between those societal forces that are able to put the necessary changes into action.

References

Adler, P. (1989), 'Capitalizing on New Product and Process Technologies: Current Problems and Emergent Trends in U.S. Industry'. Paper presented at the OECD Conference 'Technological Change as a Social Process Society, Enterprise, and the Individual'. Helsinki 11-13 November.

Alasoini, T. (1991), 'Modernization Strategies in Finnish Manufacture, Past Experiences, Prospects for the 90's', in: Kasvio et al. (1991), 39-67.

Alasoini, T. (1992), 'Toward a New Model of Rationalization and Labor Relations in the Finnish Paper Industry'. International Journal of Human Factors in Manufacturing vol. 2, 4, 355-368.

Alestalo, M., Uusitalo, H. (1986), Finland. Flora, P.(ed.), Growth to Limits. Berlin: De Gruyter.

Allardt, E. et al. (1988), Ihminen, työpaikan tärkein voimavara. Helsinki (in Finnish).

Allardt, E. (1990), 'Tieteen ja korkeakoululaitoksen kehitys keskeiseksi yhteiskunnalliseksi laitokseksi', in: Riihinen (1990), 615-633 (in Finnish).

Aoki, M. (1986), 'Horizontal vs. Vertical Information Structure of the Firm'. American Economic Review, Vol. 76, No. 5.

Asplund, R. (1991), Education, Human Capital and Economic Growth. Helsinki: ETLA (The Research Institute of the Finnish Economy), Series B No. 71.

Boyer, R.(1989), New Directions in Management Practices and Work Organization. General Principles and National Trajectories. Paper presented at the OECD Conference 'Technological Change as a Social Process Society, Enterprise, and the Individual'. Helsinki 11-13 November.

Brulin, G. (1989), Från den 'svenska modellen' till företagskorporatism? Facket och den nya företagsledningsstrategin. Lund: Studentlitteratur (in Swedish).

Bruun, N., Nielsen, R. (1992), 'The Future of the Nordic Labour Relations Model', in: Bruun et al. (1992), 221-263.

Bruun, N., Flodgren, B., Halvorsen, M., Hydén, H., Nielsen, R. (eds.) (1992), The Nordic Labour Relations Model. Labour Law and Trade Unions in the Nordic Countries - Today and Tomorrow. Aldershot: Dartmouth Publishing Company.

Cole, R.E. (1989), Strategies for Learning. Small-Group Activities in American, Japanese, and Swedish Industry. Berkeley: University of California Press.

Crouch, C. (1990), Generalized Political Exchange in Industrial Relations in Europe during the 20th Century, in: Marin, B. (ed.), Governance and Generalized Political Exchange. Self-Organizing Policy Networks in Action. Frankfurt am Main: Campus, 69-116.

Dertouzos, M.L., R.K. Lester and R.M. Solow, (1989), Made in America. Cambridge, Mass.: The MIT Press.

Dore, R. (1986), Flexible Rigidities. Stanford: Stanford University Press.

Dosi, G., C. Freeman, R. Nelson, G. Silverberg and L. Soete (eds.) (1988), Technical Change and Economic Theory. London: Pinter Publishers.

Economist (1992), 'Coming Top. A Survey on Education'. The Economist, November 21st-27th.

Edquist, C. and B.-Å. Lundvall (1989), 'Comparing Small National Systems of Innovation', mimeo.

Elam, M. and M. Börjeson (1991), 'Workplace Reform and the Stabilization of Flexible Production in Sweden', in: Jessop et al. (1991), 314-359.

Emerson, M. (1988), 'Regulation or Deregulation of the Labour Market'. European Economic Review 32, 4.

ETLA (1993), The Finnish Economy, 1993/1. Helsinki: ETLA (The Research Institute of the Finnish Economy).

European Foundation for the Improvement of Living and Working Conditions (1989), Participation in Change - New Technology and the Role of Employee Involvement. Dublin: European Foundation for the Improvement of Living and Working Conditions.

Flodgren, B. (1992), 'Co-Determination at the WorkPlace', in: Bruun et al. (1992), 46-92.

Freeman, C. (1989), Technological Change as a Social Process - Society, Enterprises, and the Individual. OECD: Paris.

Freeman, C. (1992), 'Formal Scientific and Technical Institutions in the National System of Innovation', in: Lundvall (1992), 169-187.

Handy, C. (1989), The Age of Unreason. London: Business Books Ltd.

Hertog, J.F. den and P. Schröder, (1989), Social Research for Technological Change: Lessons from National Programmes in Europe and North America. Maastricht: MERIT - Maastricht Economic Research Institute on Innovation and Technology.

Hingel, A. (1991), Archipelago Europe - Islands of Innovation. Brussels: Commission of the European Community, DG XII/H-3. Monitor/FAST, april.

Hohn, H.W. (1991), Von der Einheitsgewerkschaft zum Betriebssyndikalismus. Soziale Besschliessung im dualen System der Interessenvertretung. Berlin: Zigma.

Hyötyläinen, R., K. Toikka, R. Heinonen, A. Kasvio, J. Kiviniitty, P. Seppälä and E. Tuominen (1991), Teknis-organisatorinen muutos ja mukautuvat tuotantojärjestelmät. Tutkimussuunnitelma 14.10 (in Finnish).

Hänninen, V. and K. Kauppinen-Toropainen (1981), 'Case Studies in Job Reorganization and Job Redesign in Finland'. Institute of Occupational Health, Reviews 6, Helsinki.

d'Iribarne, P. (1989), La Logique de l'Honneur. Gestion des Entreprises et Traditions Nationales. Paris: Éditions du Seuil.

Jaikumar, R. (1986), Post-Industrial Manufacturing. Harvard Business Review, Nov.-Dec., 79-86.

Jessop, B., H. Kastendiek, K. Nielsen and A.K. Pedersen (eds.) (1991), The Politics of Flexibility. Aldershot: Edward Elgar.

Julkunen, R. (1987), Työprosessi ja pitkät aallot. Työn uusien organisaatiomuotojen synty ja yleistyminen. Jyväskylä: Vastapaino (in Finnish).

Julkunen, R., Rantalaiho, L. (1991), 'Crisis of Caring: Towards New Solutions of Welfare Services', in: Kasvio et al. (1991), 293-320.

Kairamo, K. (ed.) (1989), Education for Life: A European Strategy. Surrey: Butterworth Scientific Ltd.

Kasvio, A. (1990), Recent Work Reforms, Their Social and Political Context and the Development of Social Scientific Work Research in Finland. University of Tampere, Work Research Centre, Working Papers 7.

Kasvio, A. (1991), 'Strategies of Organizational Change in Finland', in: Kasvio et al. (1991), 189-206.

Kasvio, A., Cs. Makó and M. McDaid (eds.)(1991), Work and Social Innovations in Europe. University of Tampere, Work Research Centre, Working Papers 25.

Katzenstein, P.J. (1985), Small States in World Markets. Ithaca: Cornell University Press.

Kauppinen, T. (1989), Työelämän suhteet markkinatalousmaissa: neuvottelutoimimman kehityksen suuntaviivoja Suomessa ja 12 muussa maassa. Helsinki: Työelämän suhteiden neuvottelukunta 1/1989 (in Finnish).

Kauppinen, T. (1992), Suomen työmarkkinamallin muutos. Helsinki: Työministeriö, työpoliittinen tutkimus 30 (in Finnish).

Kevätsalo, K. (1991), Ammattiyhdistysliike ja työelämän muutos. Ammattiyhdistysliikkeen toiminta 5 työpaikan muutoksessa. Helsinki: Metallityöväen Liitto (in Finnish).

Kiviniitty, J. and T.Alasoini (1992), 'Prospects of Lean and Anthropocentric Manufacturing in the Finnish Engineering Industry'. Paper presented for the Conference Autonomy and Independent Work. Nijmegen, 30 November/1 December.

Koistinen, P. and K. Lilja (1988), 'Consensual Adaptation to New Technology: The Finnish Case', in: Hyman, R. and W. Streeck (eds.), New Technology and Industrial Relations. Oxford: Blackwell, 263-271.

Kolu, T. (1991), Työelämän laatu 1977-1990. Hyvinvoinnin koettuja muutoksia työssä. Työolokomitean mietinnön liiteselvitys. Komiteanmietintö 1991:38, Helsinki (in Finnish).

Komiteanmietintö (1991), Työolokomitean mietintö. Komiteanmietintö 1991:37. Helsinki (in Finnish).

Kortteinen, M. (1992), Suomalainen työ - tutkimus kulttuurisesta muodosta. Helsinki: Vastapaino (in Finnish).

Kosonen, P. (1991), 'Flexibilization and the Alternatives of the Nordic Welfare States', in: Jessop et al. (1991), 263-281.

Kämäräinen, P. (1991), 'Education, Vocational Training and Working Life in the Nordic Countries. Remarks on Current Educational Reforms and on the Role of Research Activities'. Finnish Work Research Bulletin,Vol. 2, 2b, 8-11.

Laine-Sveiby, K. (1991), Suomalaisuus strategiana. Helsinki: WSOY (in Finnish).

Landesmann, M. (1992), 'Industrial Policies and Social Corporatism', in: Pekkarinen et al. (1992), 242-279.

Lavikka, R. (1992), 'Teamwork Enhances Flexibility in the Finnish Clothing Industry', in: Kasvio, A. (ed.), Industry without BlueCollar Workers - Perspectives of European Clothing Industry in the 1990's. University of Tampere, Work Research Centre, Working Papers 36, 381-397.

Lehner, F. (1991), Synthesis Report: Anthropocentric Production Systems: The European Response to Advanced Manufacturing and Gobalization. IAT-FAST-APS Study Report No. 1. Gelsenkirchen: Institut für Arbeit und Technik.

Lehto, A.-M. (1992), Työelämän laatu ja tasa-arvo. Työolokomitean liiteselvitys. Komiteanmietintö 1991:39, Helsinki (in Finnish).

Lilja, R., T. Santamäki-Vuori and G. Standing (1990), Unemployment and Labour Market Flexibility: Finland. International Labour Office, Finland.

Lillrank, P., Kano, N. (1989), Continuous Improvement. Quality Control Circles in Japanese Industry. Ann Arbor: Center for Japanese Studies, The University of Michigan.

Lotti, L. (1985), 'Values Associated to Work in Finland and in Some Other Industrial Countries' in: Kasvio, A. and E. Kaukonen (eds.), Working Life in Transition. Proceedings of a Finnish-Soviet Seminar. Tampere: University of Tampere, Department of Sociology and Social Psychology, Research Reports 8, 208-214.

Lundvall, B.Å.(ed.) (1992), National Systems of Innovation. Towards a Theory of Innovation and Interactive Learning. London: Pinter Publishers.

Murray, F., (1989), 'Beyond the Negotiation of Gender: Technology Bargaining and Equal Opportunities'. Economic and Industrial Democracy, Vol. 10, 517-542.

McDaid, M. (1991), 'Convergence and Integration in Europe: A Comment on Pohjola's Paper', in: Kasvio et al. (1991), 265-289.

Nelson, R.R. (ed.) (1992), National Systems of Innovations: A Comparative Study. Oxford: Oxford University Press.

Nielsen, R. (1992), 'Protection of Employment Rights', in: Bruun et al. (1992), 93-112.

Niemelä, J. and H. Leimu (1991), 'From Industrial Conflict towards Flexible Specialization: A Case Study of Finnish Shipyards, in: Kasvio et al. (1991), 79-99.

Niitamo, O.E. (1958), Tuottavuuden kehitys Suomen teollisuudessa vuosina 1925-1952. Helsinki: Kansantaloudellisia tutkimuksia 20 (in Finnish).

Normann, G. (ed.) (1992), Mot en Ny Nordisk Välfärdsmodell. Solna: Industriförbundets Förlag AB (in Swedish).

Numminen, J. (1992), 'Tulevaisuuden teesit yliopistoille'. Yliopisto, 27, 10-12 (in Finnish).

Nye, J.S. Jr. (1990), 'Top Dog and Staying that Way'. The World in 1991. London: The Economist Publications, 14-15.

OECD (1980), Technical Change and Economic Policy. Science and Technology in a New Economic and Social Context. Paris: OECD.

OECD (1988), New Technologies in the 1990's. Towards a Socio-Economic Strategy. Paris: OECD.

Ollus, M., R. Lovio, J. Mieskonen, P. Vuorinen, J. Karko, S. Vuori and P. Ylä-Anttila (1990), Joustava tuotanto ja verkostotalouden rakenteet. Helsinki: SITRA, No. 109 (in Finnish).

Pekkarinen, J. (1988), 'Keynesianism and the Scandinavian Models of Economic Policy'. Helsinki: World Institute for Development Economics Research, Working Papers 35.

Pekkarinen, J., M. Pohjola, and R. Rowthorn (eds.) (1992), Social Corporatism - A Superior Economic System? Oxford: Blackwell.

Perez, C. (1983), Structural Change and the Assimilation of New Technologies in the Economic and Social System. Futures, Vol. 15, No. 4.

Piirainen, T. (1992), 'Hyvinvointivaltio ja yhteiskunnalliset ristiriidat: kohti postkorporatistista sosiaalipolitiikkaa?', in: Riihinen, O. (1992), 291-320 (in Finnish).

Pohjola, M. (1988), 'Työehtosopimusjärjestelmän tulevaisuus'. TTT Katsaus, 4 (in Finnish).

Pohjola, M. (1992), 'Corporatism and Wage Bargaining', in: Pekkarinen et al. (1992), 44-81.

Poikkeus, L. (1992), 'The Early Retirement as a Means of Labour Market Policies', in: Koistinen, P. et al., Finnish Labour Market Policy in Trial - Evaluation on the 1980's. Tampere: University of Tampere, Work Research Centre, Working Papers 27, 67-91.

Ranta, J. (1989), 'Tomorrow's Industries, Structural Changes and the Future of Work Organizations', in: Cooley, M., European Competitiveness in the 21st Century. Integration of Work, Culture and Technology. A Contribution to the FAST Proposal for an R&D Programme on Human Work in Advanced Technological Environments. Brussels: June.

Rehnström, P. (1992), 'Pärjääkö Suomi Euroopan tuottavuuskilpailussa'. Helsinki: Suomen työnantajain keskusliitto, teknillinen osasto 8.4 (in Finnish).

Reich, R. (1991), The Work of Nations. London: Simon & Schuster.

Riihinen, O. (ed.)(1990), Suomi 2017. Jyväskylä: Gummerus (in Finnish).

Riihinen, O. (1992), 'Näkökulmia tulevaisuuteen', in: Riihinen (ed.) (1992), 603-638 (in Finnish).

Riihinen, O. (ed.)(1992), Sosiaalipolitiikka 2017. Näkökulmia suomalaisen yhteiskunnan kehitykseen ja tulevaisuuteen. Porvoo: WSOY (in Finnish).

Roobeek, A. (1987), Crisis of Fordism and the Rise of the New Technological Paradigm. Futures, April.

Rosenberg, N. (1971), Penguin Modern Economic Readings: The Economics of Technical Change. Harmondsworth: Penguin Books.

Saari, J. (1992), Social Packages: Preconditions, Strategies, Explanations. Unpublished MA Thesis, University of Manchester.

Scott, A.J.and M. Storper (1987), 'High Technology Industry and Regional Development: a Theoretical Critique and Reconstruction'. International Social Science Journal, March.

Seppälä, E.-O. (1991), 'Kansallinen innovaatiojärjestelmä ja 1990-luvun teknologiapolitiikka'. Työelämän tutkimus vol. 2, 5, 813 (in Finnish).

Silvennoinen, H. and A. Pirilä (1992), 'Lattiatason ideaalityöntekijä - työn vaatimusten, työntekijän ominaisuuksien ja rekrytointistrategioiden yhteensovittuminen'. Työpoliittinen aikakauskirja 4, 13-25 (in Finnish).

Sorensen, B.A. (1980), 'Handlingsorientert Arbeidslivsforskning - en kort beskrivelse av tradisjoner og aktuelle forskningsprogrammer ved Arbeidspsykologisk Institutt, Arbeidsfoskningsinstituttene', in: Bjornstad, J.H., Bjaerland, T. (eds.), Arbeidsmiljo & Demokratisering. Larvik, 75-97.

Standing, G. (1988), Unemployment and Labour Market Flexibility: Sweden. Geneva: International Labour Organization.

Streeck, W. (1989), 'Skills and the Limits of Neo-Liberalism: The Enterprise of the Future as a Place of Learning'. Work, Employment and Society 3, 90-104.

Streeck, W. (1992), Social Institutions and Economic Performance. Studies on Industrial Relations in Advanced Capitalist Economies. London: Sage Publications.

Streeck, W. and Ph. Schmitter (1992), 'From National Corporatism to Transnational Pluralism: Organized Interests in the Single European Market', in: Streeck (1992), 197-231.

Thurman, J. (1991), 'Competence and Choice at Work'. On Business and Work. Towards New Frontiers. A Draft Report to be Submitted to the International Symposium Work Life in Transition, Stockholm, October 13, 1991. Uppsala: International Labour Organisation, 8-16.

Toikka, K. (1992), 'Rautahäkki: miten se rakennetaan ja puretaan'. Työelämän tutkimus vol. 3, No. 4, 35-39 (in Finnish).

Tuohinen, R. (1990), Työlle viileä sukupolvi? Nuorten työlle antamista merkityksistä, niiden tutkimisesta ja tulkinnasta. Helsinki: Työministeriö, työpoliittisia tutkimuksia 1 (in Finnish).

Valtion tiede- ja teknologianeuvosto (1991), Tiede- ja teknologiapoliittinen katsaus. Helsinki: Suomen Akatemia (in Finnish).

Vartia, P. and P. Ylä-Anttila (1992), Kansantalous 2017. Helsinki: ETLA (The Research Institute of the Finnish Economy), Series B No. 80 (in Finnish).

Vuori, S. and P. Vuorinen (1992), "Outlines of the Finnish Innovation System: The Institutional Setup and Performance", first chapter of this volume.

Vuorinen, P., 'Technology Cooperation: Do Networks Exist?', in: Kasvio et al. (1991), 131-141.

Wobbe, W. (1992), 'Anthropocentric Production Systems: Shall the EC Intervene in New Emerging Structures of European Business?' Paper presented for the Conference Autonomy and Independent Work, Nijmegen 30 November/1 December.

Womack, J.P., D.T. Jones and D. Roos (1990), The Machine That Changed the World. New York.

Ylä-Anttila, P. and R. Lovio (1991), 'Flexible Production, Industrial Networks and Company Structure - Some Scandinavian Experience', in: Kasvio et al. (1991), 103- 129.

4 Growth and Technical Change in Finland: The Role of Collective Sharing of Economic Risks

Vesa Kanniainen[1]
Department of Economics, University of Helsinki, Finland

4.1 Introduction

4.1.1 Historical Overview

The post-war economic history of Finland is rather unconventional. Partially ruined, the country lost more than a tenth of its territory in the Second World War, repopulated 400.000 new inhabitants from the ceded areas and paid 5-6% of its GDP in war remuneration over the years of the late 1940s and 2% in the early years of the 1950s.[2] This all required substantial reallocation of Finland's scarce productive capacity. Subsequently, strong policy towards economic growth built into the tax system, monetary policy, commercial policy, and the development of state-owned companies launched the economy into a fast growth over decades. As indicated by the introductory chapter of the current book, the growth of Finnish manufacturing output was 3.1%, on the average, over 1973-1990; however, this growth rate reached an even more remarkable level of 6.6% in the preceding period of 1960-1973. In international income per capita comparisons, Finland was moving up. For understandable reasons, Finland was known as the Japan of the Nordic countries in the 1980s.

The above documentation is, however, only part - the illusory one - of the whole story. Towards the end of the last decade, Finland was hit by a severe downturn with a 10% subsequent loss in her GDP over the years 1991-92, a reduction in private consumption - quite an exceptional phenomenon in western economies - and with a record-high rate of unemployment. Today, highly indebted abroad, Finland is paying about the same percentage of her GDP in interest payments on foreign borrowing as she had to pay war remuneration. The analogy to the post-war years is even more concrete and complete. The unemployment figures have reached and

---

[1] I am indebted to Matti Pohjola, Synnöve Vuori and Pekka Ylä-Anttila for many helpful comments.

[2] Cf. Ahvenainen, Pihkala and Rasila (1982).

even passed the figures of repopulation after the war.[3] At least for the time being, Finnish prosperity seems to have been a fictive illusion.

The current paper seeks to shed light on what I call the Finnish mystery. In this endeavor, I repeatedly call attention to the network of institutional arrangements which are directed towards the management of economic risks. Indeed, I claim that many of the institutions and structures which have emerged in the Finnish economy and society over the past decades have been characterized by substantial collective sharing of economic risks through the public sector. I find that the insurance through the public sector has been quite compelling, indeed. One can legitimately ask what explains this, though this question may not be easy to answer. If the answer is high risk aversion, the unavoidable associated question is whether the institutional arrangements created have truly helped to settle the problem of economic instability.

Beyond any doubt, the answer to the latter question is negative: exceptional business cycle volatility has been characteristic of the Finnish economy both in nominal and in real terms. Output and employment volatility has been quite severe. Inflation and price instability have also over long periods exceeded the corresponding figures for other OECD countries, on the average. Then the next question can be raised: has the exceptional volatility perhaps something to do with the very existence of collective risk sharing institutions? This question arises in the light of what the theory of insurance teaches us: risk insurance with high coverage may not only provide poor incentives for risk management and monitoring but may also lead to a moral hazard. We repeatedly face this problem in the current paper.

Risk management undoubtedly is essential when it comes to explain a country's "hunger" for technological progress and competition for innovations and markets and, thereby, its growth performance. Though silent on risk aspects, the modern theory of economic growth is a helpful framework for a different reason. By now, this theory has convincingly demonstrated that the growth rate of an economy over long periods can be expected to be closely related to technical progress, accumulation of human capital, quality of knowledge and a country's R&D effort to produce and to utilize new technology. To put all these linkages in a proper context, we start with a set of specific questions to be addressed in this article.

4.1.2 The Questions to be Addressed

This article raises - and seeks to answer - the following set of questions:

(i) What are the stylized facts of Finnish growth developments over 1970s and 1980s?

---

[3] I am indebted to Matti Heimonen for pointing out this analogy.

(ii) What economic models or frameworks are useful for an attempt to organize one's thinking about the relationship between economic growth and the mechanisms determining this growth?

(iii) Given the available frameworks and economic models, how is the Finnish experience to be explained? Given the network of collective insurance provided by the public sector, in what way have the managerial incentives been influenced? Given this insurance, has the economic impact been socially valuable in the normative sense?

(iv) How should Finnish technological progress and her R&D effort over the last two decades be evaluated?

When it comes to organize one's thinking of long-term economic phenomena, we thus ask what economic frameworks are available to organize the stylized facts. In the current paper, we organize our evaluation of the Finnish economy using a number of complementary economic theories. We resort to the modern theory of economic growth, for example. This theory has convincingly shown that to understand economic growth over long periods, one has to understand the determinants both of private investment and private saving. The former does not only include investment on tangible assets. Indeed, technical progress, accumulation of human capital, quality of knowledge and a country's R&D effort to produce and utilize new technology - a key intangible asset - are crucial determinants of the growth process. The theories of growth are taken up towards the later parts of the paper. In the sections to follow, we start with a description of investment and saving incentives as they have existed.

The theory of economic growth is quite aggregate and based on stylized models which are abstracted from many important institutional aspects relevant to the issues of this book. For this reason, one has also to go into the microeconomics of "the representative firm" to understand what determines innovative incentives in a market economy. Given that an R&D project is a leap into the unknown and often subject to uncontrollable economic risks, one also has to understand in what way the various methods of public R&D support participate in risk sharing. This aspect is most relevant given that the financial market provides an inefficient sharing of risks due to informational asymmetries.[4]

The limitations of the growth models, of course, stem from their aggregate nature. They are not suited for considering R&D risks or behavior under market uncertainty. They are bound to be silent on the structure of the financial system and hence on the degree of risk-sharing through the capital markets. They have not yet been extended to integrate other relevant institutional aspects like distortive factor taxes. Some branches of this literature have, however, introduced both imperfect competition and international trade. These are important elements when it comes to understand the economic growth of any particular small open economy.

[4] For this point, see Kanniainen (1992).

4.2 Some Stylized Facts of the Finnish Economy, 1970-1990

4.2.1 Macro Development and Collective Risk Insurance

1 Business Image. There is room for quite a peculiar interpretation associated with the Finnish economic developments towards prosperity. With the growth of neo-leftist social attitudes during the 1960s, business seems to have borne a negative image at least throughout the 1970s. Business was sometimes regarded as more or less "immoral". "Making money" was viewed as a secondary career choice given the more "respectable" substitutes available, like serving the country as a civil servant. One now, of course, understands the fundamental fatality of this view. However, negative attitudes towards business were powerful enough to lead, for example, to the establishment of a special organization (EVA) to counteract these attitudes and to create a more positive business image.

On the other hand, the political leadership covering also the leftist parties whenever in charge did understand the importance of a strong economy for economic welfare. Corporations, especially the export sector, have consequently enjoyed substantial political support. Over the years, this understanding developed to a dominant strategy of national consensus about the unquestionable priority of economic growth.[5]

2 Rapid Growth, High Volatility. It is a hard fact that among the western economies the average rate of output growth in Finland has been rapid by international standards. Moreover, as Vuori and Vuorinen indicate in the introductory chapter, gross fixed capital formation amounted in the 1980s to no less than 25% of GDP, on the average, in Finland, almost 5 percentage points more than the EC average and about 3.5 percentage points more than the average for small European OECD countries.[6]

Economic growth was only occasionally interrupted in the 1970s and 1980s. Even the adjustment of Finland to the oil crisis in the mid-seventies was easier than that of many other OECD economies because the Finnish terms of trade actually changed to the advantage of the country preventing any substantial reduction in real income. As a by-product, no incentives were created for structural change.

But the coin has also its other side. While capital formation in Finland has been strong, it has also been extremely volatile and unstable. Moreover, the sad fact is that a substantial part of capital formation has been directed towards the closed sector, which does not only include housing. Most disturbingly and due to strong tax incentives, fixed investment in agriculture has even exceeded that in metal and engineering industries.

---

[5] This has been called economic nationalism by Pohjola (1992).

[6] These figures are underestimates rather than overestimates. When allowing for correction in the relative prices of equipment goods, this share is increased five percentage points in Finland, cf. De Long and Summers (1991).

It is of interest to ask whether the rapid growth is policy-related and, in particular, due to exceptional technological achievements. No econometric evidence is available. Our interpretation of the Finnish data (saving, investment, interest rates, taxation), however, suggests that the growth experience has indeed been strongly policy-related.

However, in the light of the recent economic disaster of the 1990s, it is doubtful whether Finland after all has been a growth miracle. A more plausible hypothesis is that the rapid growth rate documented has been due to relatively low initial income per capita. Moreover, given the large share of annual output devoted to investment, it is more surprising that the growth rates of output and productivity have not exceeded the ones actually recorded.[7]

3 Collective Risk-Sharing: Exchange Rate Policy. High volatility of inflation and fluctuations in employment have had a close linkage with the Finnish exchange rate policy. The adjustment of the foreign exchange value of the markka has since the late 1950s followed a surprisingly regular 10-year cycle providing justification for the concept of a so called devaluation cycle.[8] The adjustments have, however, been asymmetric. Failures to revalue during booms in order to alleviate price and wage pressures have typically led to ultimate overpricing in the export market and subsequent overvaluation of the currency. In an international comparison, the devaluations of 1956, 1967 and 1977-78 have been quite substantial indeed.

It is natural to interpret the exchange rate policy as a commitment to collectively insure the risks of the shareholders of the major Finnish export sectors to restore an equilibrium between capital and labor income whenever disturbed to the disadvantage of capital income. This interpretation covers at least the period when capital flows were controlled by the central bank and when there hence was no room for speculative attacks against the markka. One can hypothesize that this type of collective insurance probably also has reduced the risk premium associated with investment in the insured sectors. This follows from the understanding that the insurance potentially has reduced the correlation between the open and the closed sectors. It is, moreover, appropriate to interpret the associated correction in the real income of wage earners as an insurance premium against an unemployment risk.

4 Average Lending Rate Regulation and Inefficiency. One of the empirical regularities of corporate finance is the close connection between annual profits and investment expenditures in spite of the existence of credit and capital markets.

---

[7] The inefficiency of Finnish production sector has earlier been discussed by Pohjola (1992).

[8] This concept was introduced by Paunio (1969).

However, there are reasons why debt actually is the desired marginal source of finance.[9] Hence, the functioning of the credit market deserves some focus.

We will put forward two claims which help to explain the overall inefficiency of investment in Finland. First, the average lending rate regulation system in the banking sector has created an adverse insurance element whereby profitable firms have had to subsidize unprofitable ones through being liable to higher than average interest rates. Second, strict limits on credit availability have promoted credit expansion to existing firms with established banking relationships as the primary criterion. Hence, there has always been the risk that some projects which would have been ranked high in NPV-comparisons have not obtained finance.

The latter problem was, however, alleviated through two mechanisms which were developed in the 1970s. First, given that firms had different access to the official credit market, a so called "grey market" developed whereby the available finance was reallocated through inter-company transactions. Its impact was perhaps not socially valuable in the sense that the terms on black market contracts generally tend to be unfavorable for the buyer. Thus, it is possible that only the high-risk projects were financed through the grey financial market.

Second, banks started to establish so called finance companies which extended business loans in packages synchronized with bank loans. The loan terms could then be better adjusted to match with the quality of the applicants. One should not, however, hasten to judge the impact on the efficiency of investment of this procedure. As always, applicants which are willing to pay an abnormally high interest rate may just signal that one is dealing with a high risk project.[10] Hence, banks unwilling to have high-risk customers cannot rely too much on the rationing effect created by high interest rates.

Regulation of the lending rates by monetary policy as part of the chosen growth strategy gave rise to a further inefficiency when combined with the inflationary pressures: real borrowing rates used to be negative. The standard theory of corporate investment then suggests that the required gross rate of return on investment has been badly distorted downwards. Having stated this conclusion, we no doubt have identified one of the major explanations behind the inefficiency of investment over the past decades in Finland.[11]

5 Monitoring and Executive Incentives. The conclusion of the previous section, i.e. that the rate of return requirement has been extremely low for investment in Finland provides the starting point for our next conclusion: the return on

---

[9] Drawing on tax debt necessitates the use of retained profits. However, tax debt aside, debt has been typically favored more or less by most OECD tax systems. Moreover, there are monitoring gains associated with the use of debt, cf. Kanniainen and Södersten (1993).

[10] For the problem of adverse selection, cf. Stiglitz and Weiss (1981).

[11] Capital markets have consequently channeled wealth from deposit holders to the investing firms and their shareholders. For these wealth transfers, cf. Kanniainen (1986).

monitoring the performance of corporate management evidently has been very low to the shareholders.[12] In periods of negative real rates of interest, the incentives of owners and the financing banks to monitor and control the management of corporations seem to be limited for the natural reason that with a small effort investment programs can be turned successful. With good reasons, one can then ask whether an inefficient monitoring also is one of the key reasons for the Finnish illusion. Moreover, it seems that the prolonged boom in the 1980s created substantially free cash flow in the sense of Jensen (1986), i.e. the cash flow left over after investing in projects with positive net present values. In the financial literature, free cash flow is associated with perquisite consumption.

Finally, growth incentives built into the tax system (see section 2.1.8) and management's tendency to overinvest have caused lock-in effects in that the dividend payout ratio has been quite low in Finland.[13] Hence, firms have apparently not been sufficiently scrutinized in the capital market in the sense of Easterbrook (1984). This all points to the conclusion that Finnish corporations have avoided the socially optimal degree of monitoring.

6 Quality of Management. Separation of ownership and control seems to be quite a recent phenomenon in Finland. Historically, a few powerful families have been in an important position in the industrialization process of the country. It is, however, not clear whether industrial competence can be inherited from one generation to another. Hence, one can ask whether professional management should have been substituted for owner-managers earlier than actually has taken place.

7 Overinvestment in the Closed Sector. Closed sector investments have competed with capital investments in the open sector and enjoyed some special public subsidies. Housing investments in particular have enjoyed public support on two accounts. First, there have been tax subsidies. Not only has the imputed rent on owner-occupied housing been tax-free. In addition, interest on housing loans has been tax-deductible, albeit up to a certain limit. Second, strict rent control has been applied to rental units with the consequence that the rental market almost disappeared with demand spillovers in owner-occupied housing. Over time, there has been a significant capitalization effect on housing prices functioning against the policy targets.

Tax shelters for investment in agriculture have been the second major distortion favoring investment in the closed sector. It is not unfair to ask whether the principle of comparative advantage ever landed on the continent of Finland.

---

[12] The model by Stulz (1990) provides some justification for this interpretation.

[13] Kanniainen (1991a).

8 Public Insurance Through Corporate Taxation. In addition to the monetary policy of negative real interest rates, with the resulting low rate of return requirement for new capital formation augmented by bailing-out property from the exchange rate mechanism, fiscal policy was built to be growth-oriented. Through the tax system, capital income has been taxed at a low effective rate both in the form of return on debt and return on equity. First, capital income in the form of interest has been tax-free for individual savers. Second, corporate dividends have been subject to a split rate system with a rather low total tax rate on distributed profits.[14] Effectively, the return on intramarginal, existing capital has been only weakly taxed. This was not changed until the end of the 1980s when Finland introduced the imputation system with the consequence that the total tax rate on distributed profits rose significantly for institutional investors setting up some tax barriers to new share issues.

Moreover, though Finland has applied the strict requirement of uniform reporting in corporate taxation[15], the tax base of undistributed profits of corporations has been subject to a number of adjustments (including accelerated depreciation, inventory adjustment, investment fund reserve, wage allowance etc.). Consequently, the effective tax rate on undistributed profits has been much below the statutory tax rate.

One implication for corporate financial structure has been that investing firms have become highly indebted. However, at the same time, their hidden reserves have also been expanded in the form of deferred taxes. These reserves have helped to maintain the debt to equity ratios under control. Hidden reserves have, however, led to the accumulation of another type of debt, i.e. tax liability to the public sector.[16]

To elaborate the arguments slightly further, it seems that the corporate tax system has actually exerted a dual impact on firms' investment behavior depending on their expectations of future profitability. For the firms which have anticipated profitable periods ahead, the tax system has provided strong investment incentives with a substantial governmental risk-sharing, while the system has preserved tax neutrality for the (more mature) firms which have stagnated with minor expansion expectations. Moreover, and associated with the theme of this book, one cannot avoid the conclusion that the monetary and tax policies, taken alone, have apparently provided poor incentives to risky R&D efforts for firms to improve efficiency and the productivity of inputs, to reduce production costs, or to innovate new products. The required low rate of return on investment in traditional capital goods and the production of traditional consumer goods can be thought to have led

---

[14] Kanniainen (1991b).

[15] Kanniainen and Södersten (1993).

[16] Due to distorted balance sheet figures on internal equity, it is rather difficult to judge the true debt-to-equity ratios for Finnish companies. Löyttyniemi (1992) has produced figures that point to the conclusion that this ratio has been between 60-70 % over 1983-1992.

to lousy managerial incentives and weak efforts to invest in something new and more risky. (See below our discussion on firms' incentives to innovate).

Extensive risk insurance through the public sector may have badly obscured, both in the minds of corporate executives and within the banking system, the idea of risks and the requirement for managing risks properly.

9 Dichotomized Exports. To understand the Finnish growth and macro performance, it is unavoidable to take account of the fact that the Finnish export sector has been strictly dichotomized. The dominating share of exports has been directed to the competing world markets. However, the share of exports to the Soviet Union reached the top of more than one fifth of exports in the 1980s and was subject to rather different incentives. First, the quality of the latter exports, most notably that of consumption goods, was of a lower standard than that of western exports. Second, the Soviet exports were sheltered from firm-specific (or idiosyncratic) risks by the public insurance.[17]

The estimated receivables associated with the former Soviet trade now amount to FIM 6-7 billion. It is thus not only that Finland lost most of her future exports to the Eastern market. She and her taxpayers also have lost their receivables on the past exports to the Soviet market.

Given the large and sheltered export market in the Soviet Union earlier, one can legitimately again ask what the impact on innovative incentives could have been. The problem was, of course, not in the diversification of exports to two different markets with different risk characteristics. The diversification turned out to entail stabilizing mechanisms.[18] The problem was that an export sector which was ex ante understood to be riskless, turned out to be most risky ex post. The public insurance could take care of the idiosyncratic risks but could do nothing to hedge against aggregate risks.

10 Socialization of Saving and Savings Risks. It has been one of the acknowledged facts that the public sector has traditionally been running surplus in Finland unlike the case in the other OECD economies. In this sense, also saving has been (at least to some extent) socialized in Finland. There have been two types of social costs of this policy. First, public saving has over decades reduced the need for private saving. Second, in periods of rising income, large public revenue has facilitated the expansion of public expenditures and public services beyond the actual capacity of the economy. The early years of the 1990s are most striking although the cycles of public expenditures have been strongly related to election years throughout the decades.

---

[17] It should be pointed out that also western exports, like exports of metal industry, have enjoyed public insurance to a variable degree.

[18] Dahlstedt (1975), Kanniainen and Mustonen (1989).

Socialization of risks used to be more or less implicit during growth periods. Today, the most notable manifestation of collective risk insurance is the crisis of the banking system. The taxpayers have guaranteed all the assets of depositors in the banking system which otherwise would largely had gone bankrupt or at least caused a dramatic contraction of loans outstanding. Together with Norway, Finland is one of the few European countries which had not even formally restricted the insurance coverage of deposits. As a side-product, full deposit insurance has made the choice of a depositor between high-risk and low-risk banks irrelevant.

11 Collective Bargaining and Risk Aversion. The interpretation of labor market mechanism in the growth process is a complex one. Growth necessitates saving, finance and sufficient capital income. Above, we pointed to the negative real rates of interest and the created wealth transfers through inflation. In Finland, there has been another key mechanism which has guaranteed high capital income (though only in the sense as an average over time). The growth of labor movement and the change in the political balance in the 1960s led to the need to attract the labor movement to support the national growth policy. National "consensus" was created as to the wage, incomes and social policies for decades.

From the national targets of promoting long-term growth, the incomes policy obviously has succeeded in guaranteeing sufficient share of capital income in the economy and high capital formation in the average sense at least.[19] Hence, there is room for a positive interpretation that the national consensus has resulted in a kind of first-best Pareto-optimal contractual equilibrium in the labor market in the spirit of cooperative actions and collective rationality. However, there is another side of the coin since any free lunches cannot exist. First, the positive impact is limited to the average rate of growth, only. National consensus badly failed in stabilizing the growth process: the Finnish economy has been disturbingly sensitive to inflationary pressures.[20] Second, the incomes policy has been possible only through extensive governmental participation resulting in expanding public services. One of the dimensions, however, has been that the labor unions have accepted high taxes on wage incomes as a premium for the extended public insurance. In a sense,

[19] See Pohjola (1992) for more on this point.

[20] Forsman and Haaparanta (1991) view foreign demand and expansion in export prices as the dominating explanation for Finnish inflation, though they recognize that the inflation in the closed sector is caused by the income effect from the open sector. One should add that the labor market mechanism is important for domestic inflation. Hence, it is possible that centralized bargaining can guarantee a lower rate of inflation and better employment than a decentralized system (Pohjola (1991), Pekkarinen, Pohjola and Rowthorn (1991)). Koskenkylä and Pekonen (1992) claim instead that wage hikes have been far too excessive with regard to productivity growth and that the average 11% increase in nominal earnings over the past 20 years has been associated with a less than 3% average increase in real incomes, only.

wage contracts have been associated with increased public insurance through social policies.

A society's degree of unionization can be viewed as a social solution to an optimization problem. For an individual agent, membership involves costs which have to be covered by the anticipated benefits. High risk aversion among potential members can be thought to be reflected in a relatively high rate of participation. By international standards, unionization of the Finnish labor force has been abnormally high.[21] This is not only a reflection of the high rate of social risk aversion so much emphasized in the current article. Union membership has also been enjoying collective support from the taxpayers in that the membership fee has been tax-deductible in personal income taxation. Moreover, the tax shelter of membership has been increased over the years together with the rise in the marginal tax rates.

Collective incomes policies have succeeded in generating sufficient capital income which also has been subject to abnormally low effective taxation. Government revenue has been mainly collected (in addition to the turnover tax) in the form of income taxes on wages and payroll taxes on labor input. Hence, incomes policies have not actually succeeded in maintaining low overall labor costs.

Throughout this section we argue that the extensive governmental participation associated with the seemingly successful growth experience actually has blurred the idea of economic risks and their proper management. As a further example we can take the insurance against employment risks: most of the associated premium is collected from the government and the employers, not from the insured. Of course, over a longer run this premium will be capitalized in nominal wage contracts. However, the incidence of the premium is fully arbitrary between industries and firms with low-unemployment risks and high-unemployment risks.

Has the unionization of society gone too far? In the cost-benefit evaluation, one has to compare the social gains in terms of increased security not only against the cost of insurance premium but in the light of the benefits and costs of the services rendered by unions. The downturn of the economy has led to problems in activities also conducted by unions.[22] From a wider perspective, one can, of course, question the rationale for tax subsidy on membership. Given the record-high and rising unemployment, this question seems today more legitimate than ever: the unions' strategy choice during the economic downturn has apparently been based on long-term targets without sufficient attention to the short-run costs to their members and society. This argument, of course, is quite an unconventional one: instead of the ordinary claim of myopic behavior it puts forward the view that the union leaders may be married to a discount rate that is too low!

12 Government Production. Presence of the public sector in production is most explicit in the case of state-owned enterprises which are giants in the national context in several areas. The leadership of these enterprises has most often been able

---

[21] Cf. Leppänen (1992).

[22] In 1992, one of the central unions (TVK) actually went bankrupt.

to operate in monopolistic conditions with no threat of bankruptcy. Moreover, also the state-owned enterprises operating in competitive international markets have been conducted without a positive probability of bankruptcy due to the taxpayers' guarantee of equity injections in the case of losses.

One has, of course, realized that through privatization the incentive structures could be improved. However, this process has started very slowly indeed.

13 Big Firms Cannot Go Bankrupt. There is one more reason why Finland has not been a market economy in the true sense of the word: big firms cannot go bankrupt. This fact derives from the inter-linkages between domestic banks and big firms. Banks' stakes in firms are concentrated and poorly diversified both as creditors and shareholders. While this gives rise to valuable monitoring gains, inefficient risk-sharing is the other side of the trade-off. The understanding has been that a failure which would threaten the stability of the banking system is out of question.[23]

4.2.2 Finnish R&D and Productivity: Evaluation

The above rather long section emphasized the extensive role of collective risk sharing in the Finnish growth process over the past few decades. We pointed out that the impact on understanding and managing risks may have been quite harmful in general and that the extensive public involvement may have distorted managerial incentives towards activities requiring less innovative efforts. This section attempts to summarize some of the stylized facts about technological change, R&D and their relationship to the economic growth in Finland. It is mainly based on the interpretation and evaluation of the material produced by Vuori and Vuorinen at the beginning of this book.

1 Technology Policy as a Risk Insurance Device. It is helpful to view a technology policy as a matter of an insurance policy whereby some risks of R&D efforts are shifted from private firms to the public sector. Insurance as such, of course, by the logical necessity substantiates risk-taking because as a result, the risks will be spread more efficiently. Risk-taking may indeed be quite productive; there will be an abnormal rate of return on a risky project that turns out to be a success. This view is confirmed by the introductory chapter of Vuori and Vuorinen.

However, whether an R&D policy functions as strongly desired is dependent upon the way it has been constructed. The principal agent theory teaches that for

---

[23] As an example, one can mention the case of Tampella. Its losses led to the transfer of ownership of its major creditor, SKOP-bank, to the possession of the central bank of Finland in 1991.

efficient effort choice and efficient risk-sharing, part of the risk has to be carried by the principal and that full insurance coverage is devastating for the incentives. Moreover, loans, grants, guarantees, equity participation and tax reliefs all have quite different effects on innovating incentives through their dissimilar impact on the way risks are shared. It is doubtful whether Finland has applied the proper policy mix and whether appropriate incentives have been created. One can refer to the limited role of venture capital and to general R&D support through the tax system. Private venture capital has not really succeeded and today is almost non-existent. Also from this perspective, risks on R&D are socialized in that the three major venture capital institutions are all public.

2 Domestic R&D or Technology Transfer? The introductory chapter by Vuori and Vuorinen provides some cross-country evidence that the Nordic countries (with the exception of Sweden) have devoted less to R&D (1 - 1.9%) over the 1970s and the 1980s than have, for example, France or the UK (2 - 2.4%) not to mention the USA, Germany or Japan (1.9 - 3%), when counted as a percentage of GDP. Essentially, this conclusion is confirmed by the figures when R&D expenditures are related to value added in manufacturing industries.

Finland also seems to have been more an importer than a producer of R&D. In addition to low patent intensity, this can be judged from Table 3.8, too. It indicates that the share of high-tech products in the Finnish exports (10.8%) is substantially smaller than is the corresponding share in the exports of many other western economies (with the exception of Norway and Denmark). In the late 1970s and 1980s, the rate of growth of R&D in real terms was very rapid in Finland (nearly 10%) when contrasted to the 6% in the OECD, on the average. Data also indicate that the Finnish total factor and labor productivity growth have been high. However, these observations may point to low initial levels rather than to exceptional performance.

Whether a strong contribution to R&D effort is optimal for a small economy is a matter of dispute. Japan is an example of a country which never has been a leader in basic research but which through applications of modern technology has gained a leading position in the product markets of many high-tech commodities. Also the Finnish R&D policy seems to have been based rather on diffusion or imports of technology than production and exports of innovations. There are thus two options to follow and the rate of diffusion chosen can be viewed as a matter of social optimization.[24]

3 One Major Development Block Only. The idea of characterizing the historical data in terms of development blocks is most helpful. For natural reasons, the forest sector has been the key block in the Finnish economy. In addition, Finland has effectively utilized the R&D of the metal and engineering industry to substantiate

[24] For this point, see Kanniainen (1992).

the comparative advantage of the forest sector. The technological innovations in this combined block have been the major element behind the strong real competitiveness of the paper industry, on one hand, and exports of technology, on the other hand, in the form of paper machines. Indeed, towards the late 1980s, Finland had become the world's leading exporter of paper machines. Also the patent statistics show that one-third of the patents were forest-related.

Has Finland had her eggs in one advanced basket only? This is not quite true: there are some other blocks. But their macroeconomic role falls much behind that of the forest block. Heavy concentration on one major development block has the disadvantage of rendering the economy highly vulnerable to fluctuations in demand for the output of this particular block. The trade-off in developing other blocks would be between more diversification, less aggregate risks in output, on one hand, but more risks in producing that block, on the other hand.

4 Skills, Education and Training. In the Nordic comparison, Finland is not a special case in formal education. In a broader international comparison, it is above the average. In the evaluation of the education system and its role one has, however, to consider a number of issues. No doubt the quality and quantity of both basic education and professional training are highly important for economic progress. However, many task-specific or firm-specific skills are acquired at the working place within the firms. Quality of this type of human capital is indirectly reflected in the real competitiveness of the industry and cannot be judged on the basis of formal education. Indeed, as the economic theory of education teaches, formal education may be more used as a signaling device by the employers reflecting the innate abilities rather than unbiased indicators of the efficiency of schooling. This is but an alternative way to pinpoint the importance of the learning-by-doing mechanism.[25] Second, as the success story of Japan indicates, the most important element in determining the productivity and efficiency at the working place is the organization of work, i.e. what incentives and penalties have been structured in the production systems. Indeed, while Finns are known to be industrious, the Finnish production sector has perhaps turned out to be much less innovative in creating motivation and developing incentives through the organization of work.

5 Innovations and Firm Size. The learning-by-doing effect also operates when considering the evolution of small firms into larger ones. The share of aggregate R&D expenditures of small firms is not very large in Finland or in Sweden. This may be so because large firms tend to have more retained profits to finance their R&D activities. Moreover, large firms can diversify some of the risks by undertaking simultaneously several uncorrelated risky R&D projects. Small firms are, however, important for total innovative activity in that many ideas tend to be produced by small firms although the commercial products may subsequently be developed

[25] See Arrow (1962b) for the hypothesis of learning-by-doing.

by larger firms. Hence, a recession which is fatal to small firms tends to have a severe adverse long-term impact on the future development of the economy. No doubt, many Finnish business ideas of the 1980s were failures and had to die out. However, in the light of the current deep recession and financial crisis in Finland, the high failure rate of small firms has a potentially substantial adverse impact on the economy over the long run.

6 Multinational Enterprises. It is one of the empirical facts that from the worldwide perspective, the bulk of the R&D effort is undertaken by large multinational firms operating at the same time in several national markets and continents. The explanation for this phenomenon can be understood with the help of the modern theory of industrial organization. It was Caves (1971) who first described the relevant mechanisms. The internationally differentiated preference and demand structures and the unpredictability of their evolution over time in various local markets is a necessary ingredient when it comes to explain the very existence of multinational firms. Acquiring and updating such information necessitates presence, costly set-up investments and continuous monitoring in local markets. It also necessitates the building-up of firm-specific assets through R&D efforts directed to process innovations to be used to produce the products best suited for different markets. Firm-specificity, in turn, gives rise to the option of monopolistic pricing to the extent products are differentiated. This all makes strategic behavior typical of the multinational enterprises.

The game is played not only between enterprises. National governments have a tendency to take strategic positions also in choosing their tax structures and the R&D subsidies so as to attract R&D investments or at least the manufacturing stage on their own territory for employment reasons.

It is, however, important to point out that imports of technology to Finland have been based on foreign direct investments to a limited extent, only. Rather, it is the case that domestically-owned firms most often have bought the know-how hence rewarding the foreign innovators by their innovations. For the social judgment of this development, one should admit that for employment reasons, foreign direct investments are not always desirable. For a positive employment effect to last, it is necessary that the production is not terminated. The acquisition of a foreign company may be part of a strategic action of creating a less competitive market in the global sense. If the acquiring firm gains the market, it may raise its global profits most by cutting down some of its plants even if the average cost is not plant-specific.[26]

---

[26] In other words, it can raise its profits by moving from a Cournot equilibrium to a contract curve with higher profits. Unfortunately, reduced competition raises prices and shutting down plants leads to unemployment. From a broader perspective, the move is anything but Pareto-improvement.

To equalize their expected rates of return domestically and abroad, Finnish firms have internationalized their production through foreign direct investments since the1970s. Many of these investments have been, however, disappointments: the profit-opportunities for latecomers tend always to be limited. To the extent they have been successes, the tax hedge for repatriation of profits from successful foreign operations has been quite high especially after the tax reform of late 1980s.[27]

Currently, the number of Finnish subsidiaries abroad is quite large. This reduces the need for additional expansion through foreign direct investments, based on equity injections from domestic parent companies. Indeed, the nucleous hypothesis of Sinn (1990) suggests that one can expect to see an internal growth of these subsidiaries in years to come.

7 Productivity Development. Technical development is anticipated to cut unit costs of production and to increase the productivity of existing capital assets and labor input if technical progress is of a disembodied type and capital malleable. To the extent technical progress is of an embodied type, there may be less room for input substitution ex post while new techniques are introduced along with new vintages of capital goods. Hence, it is of interest to ask what the time behavior of productivity looks like over a long run.

The sharp output growth of the Finnish economy has been associated with a 3.6% growth rate in labor productivity over the period 1979-90, for example (see Vuori and Vuorinen, Table 1.11) which should be compared with the much lower average OECD figure of 1.5% for the same period. Indeed, Finland beat even Japan's rate of productivity growth which was 3.0%. Moreover, Finland was not hit by the decline of productivity growth as badly as many other economies. Also the total factor productivity capturing the organizational improvements and technological advantage has grown faster in Finland (2.5%) than in the OECD, on the average (0.9%), in 1979-90 while the contribution of TFP has increased from 56% to 94%, on the average, from 1960-1973 to post 1973 period.

There are apparently several factors accounting for these observations. Finnish production has tended to be quite capital-intensive both due to tax incentives favoring investment in capital goods and to high overall labor costs. Second, one can ask whether the productivity growth should not have been even faster given the exceptionally large share of GDP devoted to capital formation. Finally and perhaps most important, one ought to keep in mind that Finland started from a relatively low level after the Second World War.

[27] Transfer pricing has helped to avoid the tax hedge associated with repatriation of profits as dividends of subsidiaries. For these issues, see Skurnik and Kanniainen (1990).

4.3 Some Lessons from the Theory of Economic Growth

4.3.1 Growth: International Evidence

To evaluate the growth performance of any country, it is interesting to have some stylized inter-country facts available before focusing attention on a single economy. The first fact seems to be that even over quite long periods, countries tend to greatly vary in terms of their growth rates.[28] For example, if the average OECD output growth over the past decades is taken as the starting point, the growth rate has recently been substantially faster in many Asian countries, including China. On the other extreme, many developing economies have produced negative growth rates. For an economist, these differences cannot be an outcome of some random process. Instead, what professional economists seek to do is to provide some consistent framework to find logical explanations in order to account for the observed facts. One such a framework is the modern theory of economic growth to be studied below.

Fortunately, there are many empirical regularities to start with. The documentation that output per worker tends to grow continuously is very strong.[29] Also many studies have identified correlations between output growth and productivity growth. Investment-to-output ratios typically seem to correlate positively with real GDP growth as do various estimates of the stock of human capital.[30] Countries that export a large share of output seem to grow faster than others[31] as do countries with slow population growth. A positive correlation has been found between the number of scientists and engineers employed in research and the growth rate of output.[32] High government consumption slows growth while high levels of government investment speed up growth. There is another regularity with fiscal policy: high marginal tax rates are associated with slow growth.[33]

In spite of these regularities, measurements are often difficult. For example, it is difficult to measure the contribution of technical progress to growth. What the recorded regularities, however, imply is that there is some hope of explaining the growth process. Growth cannot be an accident but related to economic fundamentals.

---

[28] For inter-country dispersion, see Summers and Heston (1988).

[29] Kaldor (1961), Romer (1986), Scott (1989).

[30] Romer (1989b)

[31] Romer (1989c).

[32] Romer (1989c).

[33] Koester and Kormendi (1989). Moreover, Kuznets (1988) has concluded that Japan, Taiwan and South Korea have pursued a policy of encouraging the corporate sector and of removing regulatory restrictions on business activity.

4.3.2 Theory of Economic Growth

It is clear that economic growth is related to investment, risk-taking, R&D and saving. Hence, it is related to return on saving and return on investment and the incentives to undertake risky projects. Most important, via all these mechanisms government policy towards economic growth is highly relevant when it comes to explain economic growth.

After decades' research efforts, the modern theory of economic growth has ultimately succeeded in incorporating the link missing from the earlier theories that new technologies stem from intentional actions of profit maximizing agents responding to market incentives and policy instruments. In its open economy extension, it has come to emphasize the importance of large markets for the profitability of R&D and it suggests that the comparative advantage of a nation ought to be viewed as endogenous rather than exogenous.[34] But there has been a long way to go.

For a long time, the theory of economic growth had been a prisoner of the indispensable requirement of logical consistency. Ever since Adam Smith's "The Wealth of Nations", it has been understood that competition helps to allocate productive inputs in a meaningful way and that it is the endogenous process of capital accumulation that leads to economic growth. It was not, however, until the 1980s that the economic profession gained a clearer understanding of how endogenous accumulation can be reconciled with the existence of decentralized markets. Before that, economists had to work with most unsatisfactory models where technical progress, the most important explanatory variable of economic growth, had to be considered as fully exogenous.[35]

The importance of technological progress for economic growth has, of course, never been a true issue. Growth accounting has sought to calculate the part of growth, unexplained by growth of primary factors like capital or labor input.[36] It has thus been most disturbing that the theory of growth with the competitive structure and the convexity of technological set (diminishing or constant returns to scale[37]) was fundamentally inconsistent with any explanation of how technological

[34] See Grossman and Helpman (1991).

[35] In Romer's (1989) view, the narid for growth based on endogenous accumulation was reached with the publication in 1956 of Robert Solow's and Trevor Swan's demonstration that changes in the savings rate, and implicitly in government policy variables, have no lasting effect on the rate of growth of output per worker.

[36] In Finland, the pioneering study has been Niitamo (1958). For subsequent work, see Vuori (1988).

[37] For readers outside the economic profession, it is necessary to explain the idea of convexity of technology or diminishing (contrary to constant or increasing) returns to scale. Returns are said to be of constant returns to scale variety, if an increase, say 10% of all productive factors leads to an expansion of output by precisely 10%. Returns are then increasing if the precentage increase in output exceeds 10% while they are diminishing (convex technology) if the rate of expansion falls short of 10%.

change could arise (see Romer (1989)).[38]

In contrast to the earlier neoclassical growth models, the modern models of endogenous growth have successfully demonstrated that the return on investment and the return on saving are the major determinants of the growth rate of total output. This also means that economic policy is far from being impotent as far as growth is concerned.[39]

Sala-i-Martin (1990) has shown that practically all the current models of endogenous growth actually are variants or micro-foundations of the Rebelo (1990) growth model. Many of these models incorporate the important concept of externality. In the formal sense, the new theory of economic growth emerged in the 1980's (Romer (1986), Lucas (1988)).[40] Romer (1986) assumed that the aggregate production function for an economy as a whole could exhibit non-diminishing returns on capital (knowledge, in particular) although the firm-level technologies are convex and at most allow for constant returns to scale. The explanation for increasing returns here is the positive externality that new knowledge creates for other firms. In other words, there will be technical spillovers when innovations tend to get diffused between firms. New knowledge is a kind of public good. About at the same time, Robert Lucas (1988) produced his celebrated contribution to the theory of endogenous technical growth.

It should be stated that in the paper which actually started the literature on endogenous growth, Romer (1986) followed Arrow (1962a) and Sheshinski (1967) by postulating increasing returns to scale at the economy level but constant returns at the firm level. Of course, in order to support the equilibrium with a set of competitive prices, he needed the assumption that the increasing returns are external to the firm. Moreover and, of course, again such an externality will yield a non-optimal equilibrium path. Originally, Arrow argued that the acquisition of knowledge (learning) is related to experience and he regarded investment as a good measure of increase in experience. Arrow's celebrated insight is known as the "learning-by-doing" hypothesis discussed earlier. Extensions of this model to the case of an open economy are provided by several papers, cf. Grossman and Helpman (1991).

Lucas'es (1988) model actually did not rely on increasing returns to generate persistent growth. All that is needed in his model is that there be one technology for accumulating a capital good that does not depend on any fixed factors. Lucas'es

---

[38] While convexity and price-taking behavior suggest that there cannot exist any return on innovative effort or R&D activity, the paradoxical implication is that the empirical observations of growth at the aggregate level has, by logical necessity, to arise from improving technologies. More than that, the view of diminishing returns to any single factor, developed originally by Malthus and Ricardo, strengthened the latter proposition.

[39] Later, for example Barro (1990), has linked growth to fiscal variables.

[40] The idea of externality which can be traced back to Marshall, has been part of the theory of international trade utilizing the argument of increasing returns.

model emphasizes the role of human capital in accumulation. Most important, he shows that persistent output growth is fully possible even with a constant stock of capital, provided that human capital is an essential factor in the production process and that the growth of the economy's human capital is not constrained by a fixed factor.

In Lucas'es (1988) model, human capital can change through investment: individuals will choose the amount of time they invest in their studies. All we need to generate growth is to have an incentive to invest in human capital that is nondecreasing in human capital.[41] The production function of human capital, postulated by Lucas is sufficient for the generation of endogenous growth. However, Lucas also assumed the additional mechanism based on an externality in human capital to reflect the fact that people are more productive when they are around clever people. The importance of human capital has raised the need to examine what human capital in the new growth theory means. It apparently means (Romer (1989a)) something which can be passed from one individual to the next or from a generation to the next. Apparently it contains the scientific ideas and the level of technology which have an existence outside of any individual.

The theory of endogenous growth still is very young. Subsequently, Romer (1990) has also de-emphasized the role of externality. None of the general equilibrium models existing so far provide a satisfactory description of the research and development incentives of the private sector and the way the R&D effort is compensated. Some work (Romer 1990) has been directed to abandon the assumption of price-taking. This opens up a new angle by introducing monopolistic competition. After all, the idea that the market power is essential for innovations is a very old one (see Arrow (1962a), Dasgupta and Stiglitz (1980) for the pioneering work and Reinganum (1989) for a comprehensive survey).

It has also been recognized that international trade may be a strong vehicle for economic growth in that it eliminates the need to repeat inventions and allows one to acquire them simply by exchange (Lucas (1988), Romer (1990), Grossman and Helpman (1989)).

To conclude, there are many models which emphasize R&D as an important engine of economic growth. We can think of R&D as contributing to growth in at least two ways. First, it allows to introduce new types of capital goods. The second contribution of R&D to economic growth is that it may have some spillovers on the aggregate stock of knowledge, which in turns reduces more generally the costs of producing manufacturing goods. What is needed in order to generate endogenous growth is the incentive to do R&D not to decrease over time. There are models where firms develop new varieties of consumption goods or new varieties

---

[41] As Sala-i-Martin (1990) points out, increasing returns actually are neither necessary nor sufficient to generate endogenous growth. This is good news in the sense that a number of competing hypotheses now exists to explain the growth process while a short time ago there was none.

production goods. And there are models where firms try to increase the quality of a constant number of (consumption or investment) goods. Much of this work can be found in Grossman and Helpman (1991) who also have analyzed the open economy implications and the international transmission of knowledge.

The theories of economic growth are today full of testable hypotheses. That is an avenue where more research can be anticipated in the future. For the moment and for the purposes of this article, what one can do is to borrow the intuition provided by these theories and combine it with the views expressed in our section II. Combined, this material then provides a helpful starting point for understanding the post-war economic history of Finland.

4.3.3 What Determines R&D Incentives?

What determines the creativity and motivation of innovative firms at the microeconomic level? This is a problem which has attracted plenty of theoretical interest over the past decades. Vuori and Vuorinen provide (p.2) an appropriate characterization of the issue. We elaborate that discussion only very briefly by suggesting that the key motive for R&D effort is the reward for winning the race. That reward in turn depends crucially on the industrial structure (see Reinganum (1989)).[42] One of the obstacles to R&D incentives is the appropriability problem, identified by Arrow (1962b) long time ago. The most important obstacle, however, is the problem of financing risky projects and the limits to risk-sharing.[43] These are all relevant theoretical results. No attempt will be made here to evaluate their role during the particular decades that we have been discussing.

4.4 Conclusion

In the light of the questions we posed in the introduction, what can we learn from the material in this article? We have learned that, on the average, the rate of growth has been quite rapid in Finland over the past decades. We have traced the explanation for this experience to the growth incentives created by the public sector and to the success of the forest block. However, one should not forget that in the post-war period, Finland started from quite a low level of output.

The Finnish success story of the 1980s turned into a serious economic distress in the early 1990s. Moreover, during the decades over the post-war period, Finnish growth has been very volatile and subject to repeated shocks. The material

[42] It also depends on the probabilistic return structure, cf. Kanniainen (1993b).

[43] For details, see Kanniainen (1993a).

provided by this article suggests that although the growth process has been highly policy-related in the positive sense, the other side of the coin is that the economic institutions covered by extensive public insurance of economic risks have blurred the need to face risks and manage them properly. Moreover, though public risk-sharing has promoted R&D effort, these incentives may have been offset by those public measures which have adversely influenced the structure of capital formation. As Romer (1990b) says: "applied research effort responds positively to the returns in the research sector (cf. our discussion on public risk sharing) and negatively to opportunities in other sectors' trade".

What the future policy needs as an input is perhaps a more deep evaluation of the social return - social risk trade-off of various policy measures associated with a more careful inspection of those incentive effects that the measures create inside the private sector.

References

Ahvenainen, J., E. Pihkala and V. Rasila (1982), Suomen taloushistoria 2. Teollistuva Suomi. Helsinki: Tammi.

Arrow, K. (1962a), "The Economic Implications of Learning by Doing", Review of Economic Studies 29, 155-73.

Arrow, K. (1962b), "Economic Welfare and the Allocation of Resources for Invention", in: R.Nelson (ed.), The Rate and Direction of Inventive Activity: Economic and Social Factors, NBER, Princeton University Press.

Barro, R.J. (1990), "Government Spending in a Simple Model of Endogenous Growth", Journal of Political Economy.

Caves, R.E. (1971), "International Corporations: The Industrial Economics of Foreign Investment", Economica XXXVIII, 1-27.

Dahlstedt, R. (1975), Cycles in the Finnish Trade with the Soviet Union, Helsinki: Helsinki School of Economics.

Dasgupta, P. and J.Stiglitz (1980), "Uncertainty, Industrial Structure and the Speed of R&D", Bell Journal, 11, 1-28.

De Long, J.B. and L.H. Summers (1991), "Equipment Investment and Economic Growth", Quarterly Journal of Economics, May, 445-502.

Easterbrook, F.H. (1984), "Two Agency-Cost Explanations of Dividends", American Economic Review, 74.

Forsman, P. and P.Haaparanta (1991), "Hinnat ja vaihtosuhde: onko Suomen hintataso epätasapainossa?", Paulon säätiö, 53-70 (in Finnish).

Grossman, G. and E. Helpman (1991), Innovation and Growth in the Global Economy, Cambridge: the MIT Press.

Jensen, M.C. (1986), "Agency Costs of Free Cash Flow, Corporate Finance, and Takeovers", American Economic Review, 76, 323-329.

Kaldor, N. (1961), "Capital Accumulation and Economic Growth", in: F. Lutz (ed.), The Theory of Capital, London: Macmillan.

Kanniainen, V. (1986), "On the effects of inflation: the debtor-creditor hypothesis reconsidered", The Finnish Journal of Business Economics 1986:4, 303-312.

Kanniainen, V. (1991a), "Yritysten rahoituspolitiikka: selvitys Suomen pörssiyhtiöistä 1983-1987", ETLA, (The Research Institute of the Finnish Economy), Discussion Papers No. 350 (in Finnish).

Kanniainen, V. (1991b), Yhtiöveron hyvitysjärjestelmään siirtyminen Suomessa", Helsinki: ETLA, (The Research Institute of the Finnish Economy), Series B No. 73 (in Finnish).

Kanniainen, V. and J. Mustonen (1989), "Balanced Trade Under Uncertainty: A multiperiod Non-Cooperative Game Appproach", European Journal of Political Economy, 5, 113-123.

Kanniainen, V. (1992), "Diffusion, R&D and Public Policy: Is It Better to be a First Mover or Fast Second?", in: S.Vuori and P.Ylä-Anttila (1992).

Kanniainen, V. and J. Södersten (1993a), "A New View of Corporate Taxation. Monitoring Costs, Expectations, and the Cost of Capital", Department of Economics, Uppsala University, wp 1992:2, Journal of Public Economics, forthcoming.

Kanniainen, V. (1993b), "Optimal Production of Innovations under Uncertainty", Journal of Economics, forthcoming.

Koester, R.B. and R.C. Kormendi (1989), "Taxation, Aggregate Activity and Economic Growth: Cross-Country Evidence on Some Supply Side Hypothesis", Economic Inquiry 27, 367-386.

Koskenkylä, H. and K. Pekonen (1992), "Miten palkkojen tulisi määräytyä Suomessa", KOP, Taloudellinen katsaus 1992:3 (in Finnish).

Kuznets, P.W. (1988), "An East Asian Model of Economic Development: Japan, Taiwan, and South Korea", Economic Development and Cultural Change 36 (suppl.), S11-43.

Leppänen, S. (1991), "Suomi ja Itävalta: Kaksi erilaista talouspoliittista strategiaa", TTT, Katsaus 1991:1 (in Finnish).

Lucas, R.E. (1988), "On the Mechanics of Economic Development", Journal of Monetary Economics, 22, 3-42.

Löyttyniemi, T. (1992), "Yritysten velkaantuminen ja sen ratkaisu", KOP, Taloudellinen katsaus 1992:4 (in Finnish).

Niitamo, 0. (1958), Tuottavuuden kehitys Suomen teollisuudessa vuosina 1925-1952, Kansantaloudellisia tutkimuksia 20, Helsinki (in Finnish).

Paunio, J. (1969), "Comments on the Papers of Goran Ohlin and Andre Marchal", in: P.Samuelson, (ed.), Proceedings of the Third Congress of the International Economic Association, Macmillan, St.Martin's Press, 221-226.

Pekkarinen, J., M. Pohjola, and B. Rowthorn (1991), (ed.): Social Corporatism - A Superior Economic System?, Oxford: Oxford University Press.

Pohjola, M. (1991), "Tulopoliittinen neuvottelujärjestelmä ja hintataso", Paulon säätiö, 73-87 (in Finnish).

Pohjola, M. (1992), "Millaista Suomea olemme nyt tekemässä?", Uudistuuko vai rapistuuko Suomi - hyvinvointiyhteiskunnan uudet valinnat, 27.8.1992 pidetyn seminaarin alustukset (in Finnish).

Rebelo, S. (1990), "Long Run Policy Analysis and Long Run Growth", NBER wp, April.

Reinganum, J.F. (1989), "The Timing of Innovation: Research, Development, and Diffusion", R. Schmalensee and R.D. Willig (eds.), in: Handbook of Industrial Organization, vol I, Elsevier Science Publishers B.V.

Romer, P. (1986), "Increasing Returns and Long-Run Growth", Journal of Political Economy, 94, 1002-37.

Romer, P.M. (1989a), "Increasing Returns and New Development in the Theory of Growth", mimeo.

Romer, P.M. (1989b), "Human Capital and Growth: Theory and Evidence", NBER wp No. 3173.

Romer P.M. (1989c), "What Determines the Rate of Growth and Technical Change?", The World Bank, wp No. 279.

Romer, P.M. (1990a), "Endogenous Technological Change", Journal of Political Economy, 98, 71-101.

Romer, P.M. (1990b), "Capital, Labor, and Productivity", Brooking Papers on Economic Activity, 337-367.

Sala-i-Martin, X. (1990), "Lecture Notes on Economic Growth (II): Five Propotype Models of Endogenous Growth", NBER wp 3564.

Scott, M.G. (1989), A New View of Economic Growth, Oxford: Oxford University Press.

Sheshinski, E. (1967), "Optimal Accumulation with Learning by Doing", in: K. Shell (ed.), Essays on the Theory of Optimal Economic Growth, M.I.T. Press.

Sinn, H.-W. (1990), "The Vanishing Harberger Triangle", Journal of Public Economics.

Skurnik, H. and V. Kanniainen (1990), "Ulkomaisten osinkojen verotuskohtelu yhtiöveron hyvitysjärjestelmässä", KOP, Taloudellinen katsaus (in Finnish).

Solow, R. (1956), "A Contribution to the Theory of Economic Growth", The Quarterly Journal of Economics, 70, 65-94.

Stiglitz, J.E. and A.Weiss (1981), "Credit Rationing in Markets With Imperfect Information", American Economic Review, 71, 393-410.

Stulz, R.M. (1990), "Managerial discretion and optimal financing policies", Journal of Financial Economics 26, 3-27.

Summers, R. and A. Heston (1988), "A New Set of International Comparisons of Real Product and Price Levels: Estimates of 130 Countries", The Review of Income and Wealth 34, 1-25.

Swan, T.W. (1956), "Economic Growth and Capital Accumulation", Economic Record, 32, 334-61.

Vuori, S. (1988), "Total Factor Productivity and R&D in Finnish, Swedish and Norwegian Manufacturing Industries, 1964-1983", ETLA (The Research Institute of the Finnish Economy), Discussion Papers No. 257, 1988.

Vuori, S. and P. Vuorinen (1993), "Outlines of the Finnish Innovation System: The Institutional Setup and Performance", first chapter of this volume.

Vuori, S. and P. Ylä-Anttila, (eds.) (1992), Mastering Technology Diffusion - The Finnish Experience, Helsinki: ETLA (The Research Institute of the Finnish Economy), Series B No. 82.

5 The Role of Firms in Technological Change - A Company View

Yrjö Pessi
Former President of Kemira Group, Helsinki, Finland

5.1 Introduction

Some important background factors were instrumental in the shaping up of the industrial structures in Finland. The wood processing industry, the central sector of our industry, was founded on the basis and availability of domestic raw material. The first years of our national independence saw the beginnings of our basic chemical industry to serve the needs of the young nation and somewhat later the birth of the oil refining and petrochemical industries. Finnish industry developed actively and largely in response to domestic demand and after WWII also to shoulder and carry through the required war reparations.

The 1970s marked a watershed in Finland's industrial development. As a result of the global oil crisis towards the middle of the decade world market demand declined while many sectors of Finnish industry had by then brought up their output capacity to cover domestic demand. Some basic heavy industries were facing dwindling prospects while in industrially developed countries high technology industries were attracting increasing attention. Consequently, many companies reached the point of a serious and timely reassessment of operational strategies.

By the early 1970s many Finnish companies had accrued a considerable amount of internationally comparable technology and expertise. In the evening out of domestic market demand internationalization became an avenue to put the expertise to good use, a practice already familiar e.g. among forest industry concerns. Furthermore, increasing international competition was clearly in evidence. Big companies did realize the trend and sought to link their international activities as an integral part to their operating strategies.

The following is a review of the factors, some of them already acknowledged and applied by industries, others general in nature, which have been and still continue to be determinative of Finland's industrial development and the innovational activities of industrial firms. The understanding and views presented here have grown out of involvement and work in industry and are based mostly on the experience, knowledge, and observations thereof without any express attempt at scientific scrutiny.

The term innovation, as used in the present context encompassing our innovation system and its impact on structural changes in industry, signifies a process which results in an idea's or an invention's becoming economically useful or otherwise applicable (Kanerva et al. 1989, p. 41).

5.2 Ownership and Company

Investments constitute the most important decisions in a company and are always propelled by innovative activities regardless of whether the investments are directed to tangible or intangible ends. The expansion of existing output capacity is almost invariably realized through development of technology. As a result, investment in R&D constitutes an important part of innovations and involves and channels the owners' decision making when company strategies come up for review.

Presently no data based on types of company ownership is available to enable us to chart investments in R&D and working capital, except that concerning working capital investments in government-majority companies. This group has invested far more than their share of the value of industrial finished goods would warrant (The Council of Government Firms 1992, p. 9).

Also the overseas investments of the government-majority companies have been considerable. The share of the overseas investments of Finnish companies ranged between 20 to 52% in the period of 1984 through 1991 (unpublished data). With the saturated domestic market, companies have come to appreciate the opportunities for expansion offered by overseas involvement.

The vitality of a company's innovational drive is also determined by the attitudes of the shareholders and their executive appointees to the proposed growth objectives and the extent of the R&D program. On average, small-scale family businesses tend to be cautious about the benefits of long-term R&D activity and frequently refrain from expanding their operations if their current size and effort already satisfy their expectations.

As investors and co-owners, banks have in recent years played an active role in managing company mergers, argued for on the basis of attendant synergetic advantages and effective structural reorganization. However, interests other than those of the industries involved have been in evidence in the background. With emerging problems of profitability, companies have sought to incorporate and also adhere to the interests of the investors. To all appearances, the solutions found seem in most cases to have been satisfactory.

The government-majority concerns are almost invariably capital-intensive basic industries, governed not only by an executive board but also an advisory council. During the past few decades these concerns have been vigorously developing their lines of industry, concentrating on their existing production but also spawning new industries. Their R&D activity has been fairly consequential. The owner base of these companies has provided management with a firm foundation for sustained planning. However, it has not been able to secure enough venture capital to help keep up with investments, and that is why the companies' own capital resources have remained below those of their competitors. During periods of economic slowdown the capital deficiency translates into meager investments. Since the early 1980s, the government-majority companies' eagerly embraced internationalization process has significantly affected their adoption of innovations.

The government resolution to gradually privatize the government-majority companies will in the next ten years widen the scope for ownership and structural rearrangements. Our industry will also benefit from increasingly less restricted foreign ownership. In recent years the experiences from international ownership have been encouraging.

Cooperative companies operate mainly in agriculture and the food processing industry and are naturally geared to the domestic market. Some, e.g. Keskusosuusliike Valio [The Central Cooperative Valio], have invested considerably in research and product development, including basic research, and their product innovations have resulted in a number of high quality products. In this industrial sector, too, the owner base has provided management with optimum conditions for long-term planning. Owing to the self-contained nature of the sector, the challenge to innovations from international competition has been evident only in the last ten years.

In terms of innovation projects the most unfavorable form of ownership is two-partner joint ownership, for when the owners' interests clash, even the company's articles of association may contain limitations to operations. A good example here is Fermion Oy.

Joint ownership between a private and a government-majority company (Pekema Oy and Stymer Oy) creates an unusual owner base to develop operations. This has been the model in setting up new industries in Finland when only one owner has supplied the main raw material. These companies have gradually been absorbed by Neste Oy.

Developing a new research-intensive industry has been equally unsuccessful on the basis of several owners. An example here is the biotechnology firm Genesit Oy, which went out of business in ten years. Because of the wide compass of the technology and the perseverance it required, the company failed to satisfy the expectations of all of its owners who finally lost interest in repeated capital investments.

To summarize the significance of company owner base for innovations, it is generally more favorable and advisable to have one sufficiently strong and dominant center in the base as an underpinning for long-term planning and realization of projects. In this regard the situation in Finland has been fairly good over the past few decades.

5.3 Company Culture

In time each organization develops its intrinsic culture, which in part determines staff behavior. Consequently, the management patterns and executive styles in an industrial company become significant indicators of the company's, its departments', and work units' culture. We should not forget the input of union activity and the company shop steward system either. Generally speaking, the company culture within our industries has instilled confidence and vigor into entrepreneurial activity and our organizational model has proven itself viable.

Traditionally, company culture harbors certain weaknesses, too. As the company becomes older and more settled, it accrues a certain inertia in its functioning, which could be described as resistance to change or closing one's eyes to new trends in the environment. Such oversight can affect both staff and owners. The phenomenon may play a part in that some former heavily industrialized regions, such as the Tampere area, have in the past two decades lost some of their industries. Some companies' innovation systems betrayed weak links, and others reacted slowly or ineffectually to new patterns and changes in their environment.

In recent years, an organizational change whereby business activities have been diversified and set up as separate companies has taken a firm foothold and brought organizational flexibility in its wake. In fact, the importance of innovations fueling change in organizations is on the rise for many reasons. For one thing, companies' extramural environment is in constant flux. Next, to make the most of the increasing skills and competence of staff necessitates regular checking and updating of work organization and staff assignments. Furthermore, the shift in market demand from mass to tailored products and the underlying and enabling technical advances thereof all foster the adoption of flexible production models (Ollus et al. 1990, p. 13), which in turn revitalize and refashion company culture.

5.4 Cooperation for Information Technology Diffusion

Cooperation among industries can take several forms. Two or more companies together can realize a development project. An industrial company can utilize the capacity and resources of a research institute or a university in a joint research project. A development project can also be carried out among several interested parties, e.g. a firm in cooperation with research institutes, universities, and investors. Here especially the Technology Development Centre (TEKES) has been admirably forward to promote such development projects (Tomner 1990, p. 21). In this context one should also mention the technology parks which have been springing up in various parts of the country.

Finnish firms have also joined international research programs, particularly the EUREKA projects. Cooperation such as this will but increase in the future, for an ever greater share of Finnish research funding will be channeled to European research initiatives.

It is fair to point out though that the opportunities opened up in cooperation have not always been fully exploited. In the 1970s the negative attitude adopted by the government to cooperation between industry and universities turned out very debilitating. Such attitudes have now vanished. It is essential to promote readiness, through attitudes or otherwise, for international research cooperation within companies as well as research institutes.

In some cases big companies have sought to upgrade the quality of teaching and research in the universities through industry-sponsored "academy projects," in which also research staff from industries have participated. Such projects would be even more welcome now to further cooperation and to promote interest in innovations.

Despite the fact that the diffusion of knowledge is a demanding undertaking and that cooperation has not reached its most fruitful forms, Finnish industry has so far been very successful in maintaining its level of competence. All this has been facilitated by companies' concentration on the R&D in their own fields, though largely in conventional ways. In terms of the national economy, the need to diversify the industrial base has become an important issue with a special emphasis on joining forces in cooperation.

5.5 Some Business Strategies

Traditionally Finnish mass production-oriented process industries, such as paper and pulp, mining, and in part also the chemical industry, have centered on manufacturing and relied on the availability of raw materials. Particularly the chemical industry's emphasis on production shows in the integrated refining of the industry's by-products. One could mention here e.g. the processing chain whereby the environmentally harmful sulfur from the mining industry is first refined into sulfuric acid, the acid then used to produce, among other things, potassium sulfate or sodium sulfate, and the chlorine from the latter process refined further into calcium chloride. Only in the past decade have the process-oriented business practices been harnessed to increasingly market-oriented thinking with more emphasis on the use and development of products. Titanium oxide manufacturing, on the other hand, exemplifies a strategy in which marketing dominates right from the first and in which the quality of the product determines its competitive edge on the export market.

New high-tech companies face their biggest problems usually in marketing, for they are all too often set up almost exclusively on the basis of technical expertise (Autio et. al., 1989, p. 17).

Finnish firms have applied various approaches to initiating new manufacturing and business practices. The most usual practice in starting out has been to purchase know-how and then gradually and increasingly rely on the company's own accumulated know-how and research effort when they have reached a firm footing. As a result, the past two decades have witnessed the introduction of several new processes within the chemical industry, such as the enrichment of phosphate ore, the manufacture of hydrogen peroxide, the catalytic conversion of vehicle emissions, and the mica and gypsum pigments. Valmet Paper Machinery Inc. is the world leader in the design of paper making machinery.

Another useful expedient to gain needed technology has been to become a shareholder in and contract one's own researchers in to a leading enterprise in the field, as Kemira Oy has done in genetic engineering technology. Temporary recruitment of foreign specialists for company research programs has also been tried and will most likely be widely adopted in future.

Environmental protection has of late had its impact on ways of conducting business. In developing its technologies Finnish industry has produced a great deal of knowledge applicable to technological solutions for maintaining and improving the environment. Technologies are also being developed for the express purpose of solving environmental problems. However, such activity has so far had no appreciable economic benefits for companies. Yet the effort in the field has its interest and relevance especially in view of the severity of the environmental problems looming all over the world.

5.6 Company Development and the Market

Some market factors have greatly facilitated the business of Finnish industries, e.g. the rather long-standing assured domestic demand, the dominant market position of several fairly large companies, and the clearing-based counter trade with the former Soviet Union.

During high demand and shortage of goods, the expansion of companies is keenly focused on output and processing methods. So was it also widely in Finland in the 1950s and 60s. That served a purpose, for though companies showed little incentive to market-oriented product development, they gained strength technologically and economically and that again prepared the ground for them to face the challenge of domestic and overseas competition.

In their inception some major Finnish companies had their main line of industry protected or otherwise singled out for decades. In terms of development this had the positive, often ignored, effect of enabling the companies to mature under such protection into internationally top technology concerns. Today these companies, widely international, operate in a competitive free market economy. The course of development has thus been good for our industry and hence for the national economy.

As regards Finland's trade relations, the counter trade with the Soviet Union, which lasted until the late 1980s, was so extensive in volume as to steer the interests of many industries in Finland. However, product design and competitive pricing often lacked behind western market standards, and all too often product and process development as well as marketing plans for the west were largely ignored, e.g. by many firms in the clothing, textile, and footwear industries. This happened to coincide with a partly training-induced decline in the quality of designer goods, while in furniture manufacturing design succumbed to automated mass production. In many cases the overtly dominant market position of one client risks a company's capabilities to maintain steady progress.

The domestic market is seldom wide enough for companies, so they have to be capable of marketing their products overseas. Small companies are often the first to have problems with exporting, because especially in starting out they tend to lack sufficient resources and skills. These hurdles will be significantly lowered by the support initiated and offered to small company marketing rings by the Ministry of Trade and Industry. Alliance with a foreign company is also an option to secure international markets and the synergetic advantages involved in development and manufacturing. Some such partnerships of recent years have so far been positive. Through acquisitions or alliances, companies have often managed to strengthen their market positions and increase their talent pool and thus guarantee resourceful development performance. This has enabled them to focus more keenly on specialization, a trend which is expected to continue.

5.7 Management and R&D

The R&D which aims at generating new business is a long-term investment commitment, and management's attitudes to its own R&D effort do have a bearing on the innovations achieved.

It has been calculated that to keep up the present standard of well-being in Finland requires a one and a half time increase in current industrial output (The Prime Minister's Office 1992). This calls for a diversification of our industrial base, which call has been positively answered by the growth and progress of our high-tech industry (Helsingin Sanomat 1992). Its 1992 exports reached already 15% of the total (unpublished data). The industry has received support from e.g. TEKES (The Technology Development Centre) and SITRA (The Finnish National Fund for Research and Development). TEKES has so far participated in 820 ventures, 440 of them successful. According to the State Audit Office this created 5000 new jobs and FIM 23 billion in export revenue. By the end of 1991 SITRA had invested in 34 growing high-tech based companies and in 10 strategic ventures (SITRA 1991). In addition, SITRA can boast 25 years' experience in other venture capital investments and related activities. Another important domain of its operations is to help create cooperation networks.

The advances of our high-tech industry prove that even demanding innovations have not been impossible. Accordingly, some of our basic industries should redouble their R&D effort, the initial link of the innovation chain. Renewed research initiatives are necessary e.g. to refine and upgrade processes for further utilization of wood and its components. A similar commitment to research, preferably shared by companies, is needed in order to raise the value added content of the products of the food manufacturing industry. One could pose the question whether indeed the management of large corporations should nurture new incipient R&D activity by investing in R&D personnel (cf. Klus 1985, p. 44). Should corporate management create closer ties with e.g. universities to make the most of any future research cooperation (cf. Kanerva 1989, p. 82)?

The situation is, of course, somewhat different in small progressive companies. Researchers who found companies and entrepreneurs who develop new products are truly active innovators and developers (Kanerva 1989, p. 42), but here obstacles often appear towards the end of the innovation chain. Their management skills may be inadequate or they may lack needed capabilities and resources to create market channels. These small businesses have frequently been supported by risk investors to promote innovations and, when needed, to instruct in managerial skills.

In most cases it takes more than ten years to bring a new business venture from idea to operation. An example of a time-consuming project successfully completed is xylitol manufactured by the present Cultor Oy. On the other hand, Lohja Oy, the developer of the electroluminescent display, considered it more worthwhile to sell the project before becoming a full-fledged venture, remaining, however, on the project. A biological fungus control system, the Mycostop project by Kemira Oy, is nearing completion after more than ten years of work.

Management attitudes to R&D vary from one industrial sector to another. The mechanical wood processing industry has so far paid only fleeting attention to product development. Mechanical engineering, on the other hand, applies various technologies and deems it essential to invest in technology transfers and the knowledge to be gained therefrom.

In many companies management has been positive about the research in biotechnology but has been slow to reap the full benefits of the top quality work done in many research institutes. One reason here is the lack of cooperation between universities and industries.

The food processing industry needs to stimulate its R&D work, for it is necessary to upgrade the processing of the bio-based raw materials produced in fields and woods. So far the industry has confined itself largely to conventional bulk products.

The forest industry has a common research facility, the Central Laboratory. However, organizing research on this basis tends to undermine management's close links with ongoing R&D. A case in point is the current cellulose manufacturing process which dates from the previous century. A new process, tested feasible at the pilot level, is now being developed (Tirkkonen 1992).

The paper and pulp industry has shown interest in product development for less than ten years owing to the fact that the technology had earlier been developed mainly by machinery manufacturers. Finland is now the world leader in paper mill technology.

Management has been actively supportive of R&D, e.g. in the pharmaceutical, petrochemical, and chemical industries as well as in measurement, electronics, and telecommunications technologies.

The road ahead for Finnish industry is definitely towards more sophisticated processing and advanced expertise where management attitudes to R&D play a key role. The necessity to combine the efforts of several technologies is already a fact in many projects and places a particular emphasis on cooperation and the capacity to utilize existing knowledge. Various joint technology programs are among the incentives to promote cooperation (The Ministry of Trade and Industry 1990).

5.8 The Importance of Becoming International

The overseas market orientation of Finnish companies accelerated especially in the 1970s, followed by a production-wise internationalization in the 1980s (Reinikainen 1991). All this was natural follow-up to industrial growth and accumulation of know-how. The internationalizing trend has brought along incentives to cost cutting in manufacturing and to improvement of product quality. Acquiring companies has not only added to plant and facility holdings but also opened up markets, which has often been the main motive to buy. Internationalization has thus contributed to company expertise, helped focus R&D activity along company product lines, and opened channels through subsidiaries to participate in the host country's national research programs.

These are, however, not the only reasons for Finnish companies' push overseas. The opportunities unfolding in European integration and global internationalization have been important background considerations.

By acquiring plants in various countries chemical process industries in particular have been able to optimize the flow of their products and raw materials. Consequently, process control has become more flexible and customer service has been guaranteed even in unexpected circumstances.

An overseas company buy may also be strategically motivated by acquisition of an otherwise unavailable process. An example of this is the Kemira Oy acquisition in the U.S. which primarily aimed at taking over the chloride process-based titanium dioxide technology and production.

Besides Europe, Finnish companies have extended their operations to America and Asia, and an increasing number of them have reached the point when global operational strategies have become the order of the day (Maula 1991).

The rise of industrial output in low income countries has affected Finland's industrial structure and manufacturing organization. Consequently, Finnish companies have already moved or are in the process of moving their operations into these countries, the clothing industry being a good example (Vuori and Ylä-Anttila 1989, p. 80). The now independent Baltic states are providing opportunities for regional subcontracts, of which Finnish companies have already begun taking advantage.

In general, experiences from internationalization have been positive and have revealed the pitfalls to be avoided. One can also justifiably say that without the internationalizing drive of Finnish companies during the past twenty some years several industries would now be facing ever slimmer prospects or even extinction. Internationalization has been the way to invigorate and renew industry.

5.9 Venture Capital Investment

An important part of the innovation system is venture capital investment. In 1990, Finland had seventeen organizations targeting small and medium sized companies with fixed-term minority-share investments, thus actively developing their targets. The number of venture capital investors has now shrunk to about ten. Total investments in companies have been just short of FIM 200 million (The Finnish Venture Capital Investors' Association 1991), and the recipients have been typically companies specializing in industrial manufacturing and customer services. Technology intensiveness has been well on the increase and the recipients include a number of high-tech firms. Venture capital investors offer not only capital but also expertise and know-how.

The Finnish National Fund for Research and Development (SITRA) is a major venture capital investor in Finland. The Regional Development Fund (KERA), too, acts as a venture capital organization. These two together contribute more than half the venture capital investments in Finland. Besides venture investments, SITRA has also been active in creating contact networks both at home and abroad, an activity which has been of practical import in information diffusion and interaction between organizations.

In recent years, investors' total venture capital and investments have been steadily growing, while companies' own venture capital remains one of the weaknesses of our innovation system. Removal of this bottleneck would effectively set the stage for diversified and growth-oriented industrial output.

5.10 General Conditions for Industrial Activity

Energy costs have been the most competitive part of all production costs for Finnish industry. The raw material costs more to the wood processing industry than to its competitors. Capital costs are also higher in Finland than among its closest trade rivals, and total labor costs have by now reached western levels. Yet effective work time is among the shortest in Europe, cutting back the gain on the capital invested in production facilities. Hence to remain competitive companies have had to compensate for disadvantages with gains somewhere else.

Finnish industry has not been able to strengthen its capital structure to equal that of its international rivals. The share of its equity capital is lower than in European, American, and Japanese companies, caused by such factors as corporate taxation, increased wage and labor costs, and the high cost of raw material in the forest industry. This leads into Finnish companies' lowered investment capacity during periods of economic decline.

Increasing attention will in future be paid to the locations and infrastructure of industries, for public opinion and environmental considerations will inevitably be a part of the decision making. There is already proof that it is virtually impossible to set up heavy industry in new locations because of adverse public pressure. Internationalization has introduced overseas investments, which may now curtail investments at home. The conditions for industrial activity in Finland will also affect the amount of foreign interest and investment in the country. Consequently, with the increase of options, the country's economic policies and public attitudes will be of crucial importance.

5.11 Conclusions

The following conclusions can be drawn from the development and structural change of Finnish industry and the working of the industry's innovation system over the past twenty years:

- In general, our company culture has strengthened the above developments. The capacity and readiness to quickly respond to change as well as the demand for innovations presuppose a corresponding evolution in company culture.
- The owner base of industrial companies has on the whole been favorable and conducive to long-term planning but not adequately supportive of venture capital growth relative to company expansion. In setting up new businesses venture capital investment constitutes one of the weaknesses of the innovation system; such investment efforts should be backed up and boosted.

- Within the framework of existing taxation practices, wage and labor costs, and the cost of raw material in the forest industry, companies have not had the leeway to accumulate their equity capital to the level of their foreign competitors. Broadening the owner base with overseas holdings and privatizing government-majority concerns will minimize the above disadvantage.
- As it is important in terms of the Finnish national economy to expand and diversify industries, it is equally important to be able to transfer technology from research to companies. All the various possibilities for cooperation in this regard have not yet been fully exploited. Hence it is important and necessary to promote the startup and performance of cooperation networks.
- Finnish companies have demonstrated commendable technical performance and manufacturing-oriented business acumen while their market orientation has been dampened by the dominant or even protected market position of many of them at home and the clearing-based counter trade with the former Soviet Union. The internationalization of production, gaining ground especially in the 1980s, will reinforce companies' market orientation and enhance their readiness to function in an integrated Europe.
- Exporting high technology is rapidly increasing and can best be guaranteed continuance by encouraging R&D and promoting opportunities for venture capital investment.

References

Autio, E. et al., (1989). Tutkimuksen tehokkaampaan hyödyntämiseen [More out of Research]. Helsinki: SITRA, report no. 104 (In Finnish).

Kaarre, M., (1992). Kansainvälinen tietoliikenne suomalaisten yritysten toiminnassa [The Use of International Telecommunications in Finnish Companies]. Helsinki: SITRA, report no. 120 (In Finnish).

Kanerva, R. et. al., (1989). Ideasta innovaatioksi [From Idea to Innovation]. Helsinki: SITRA, report no. 102 (In Finnish).

Kauppa- ja teollisuusministeriö, 1990. Teknologiaohjelmatoiminnan linjat 1990-luvulle. Komiteamietintö 1990:2. [The Ministry of Trade and Industry, 1990. The Outlines of the Technology Program for the 1990s. Committee report 1990:2] (In Finnish).

Keinälä, S., (1990). Finnish High-Tech Industries and European Integration. Helsinki: ETLA, C 60.

Kemian keskusliitto r.y.,(1986). Suomen kemian teollisuus. [The Central Union of Finnish Chemical Industries: The Finnish Chemical Industry]. Mikkeli (In Finnish).

Klus, J.P. and project group, (1985). Effective Technology Transfer. Project for Developing Continuing Engineering Education in Finland. Helsinki.

Maula, M., (1991). Kansainvälistyminen ja tietotekniikka. [Internationalization and Information Technology]. Helsinki: SITRA, report no. 115 (In Finnish).

Ollus, M. et. al., (1990). Joustava tuotanto ja verkostotalous. [Flexible Production and Network Economy]. Helsinki: SITRA, report no. 109 (In Finnish).

Opetusministeriö, 1992. Korkeakoulut ja rakennemuutos-työryhmän muistio, 1992:33. [The Ministry of Education: The Committee on Universities and Structural Change, a Report] (In Finnish).

Reinikainen, V., (1991). Yritysten kansainvälistymisen teoria ja syvenevän integraation haaste. Helsinki: ETLA B 70. [The Theory of Corporate Internationalization and the Challenge to Further Integration] (In Finnish).

SITRA, 1992. SITRA (1991), vuosikertomus [annual report] (In Finnish).

Suomen riskisijoitusyhdistys, (1991). Venture tiedote 2/91. [The Finnish Venture Investors' Association: Venture News] (In Finnish).

Teollisuuden koulutusvaliokunta, (1992). Koulutuksella tulevaisuuteen. Report A/13/1992. [The Finnish Industrial Training Commission: Education for the Future] (In Finnish).

Tirkkonen, M.L., (1992). Milox-menetelmällä pisimmälle Suomessa. Kemira. 3/1992. [Milox Makes It Big in Finland] (In Finnish).

Tomner, S., (1990). Rapport över utvärdering av Tekes teknologiprogram. Handels- och industriministeriet, 1990 (In Swedish).

Valtioneuvoston kanslia, (1992). Suomi tarvitsee uusteollistamisohjelman. Valtioneuvoston kanslian julkaisusarja 1991/1. [The Program for Finland's Industrial Renewal; The Prime Minister's Office Publications] (In Finnish).

Valtionyhtiöiden neuvottelukunta, (1992). Valtion yhtiöt 1991. Helsinki. [The Commission of Government Firms: Government Firms] (In Finnish).

Vuori, S. and Ylä-Anttila, P., (1989). Vaatetusteollisuus tienhaarassa. [The Clothing Industry at the Crossroads]. Helsinki: SITRA, Series A no. 93 (In Finnish).

6 Comparing National Systems of Innovation. The Case of Finland, Denmark and Sweden

Birgitte Gregersen, Bjørn Johnson and Arne Kristensen
Institute for Production, University of Aalborg, Denmark

One way to get a better understanding of innovation processes in a specific country might be to carry out a systematic comparison with other countries. In this way the international diversity of modes and methods of innovation might be used to enhance the comprehension of national innovation possibilities. The concept of 'national systems of innovation' might be useful as a tool in such comparative studies.

By a national system of innovation we mean all parts and aspects of the economic structure and the institutional set-up of a nation which affect the innovation process (Lundvall 1992). This definition is motivated in the following way: We regard innovation as a ubiquitous phenomenon in the modern economy. In most parts of it we can find on-going innovation activities of different kinds. Sometimes these result in radical breaks with the past, making important parts of the accumulated economic knowledge obsolete. Often, however, innovation is gradual and cumulative, present innovations being, in some respects, a continuation of past ones. The distinction, often made in innovation theory, between invention, innovation and diffusion as separate stages now becomes blurred, and innovations appears not primarily as single events, but more as a process.

We also regard innovation as ultimately resulting from different kinds of learning processes through which new knowledge emerges, or pieces of existing knowledge are combined and put to use in the economy. Furthermore, most forms of learning, except maybe simple imprinting, may be regarded as an interactive process in which people put different pieces of knowledge together into something new. It follows from this that innovation is rooted in the institutional set-up of the economy, since institutions may be defined as the sets of habits, routines, norms and laws that regulate the relations between people and thus shape human interaction and learning (Johnson 1992).

Since learning partly emanates from routine activities in economic production, innovation must also be rooted in the prevailing economic structure. Different technological bottlenecks and opportunities, income elasticities and linkages between industries make learning in some industries and periods much easier than in other industries and periods (Andersen 1992).

So, innovation processes can fundamentally be seen as resulting from interactive learning processes that are affected by both the economic structure and the institutional set-up of the economy. Since both the economic structure and the institutional set-up differ between nations we may, thus, define national systems of innovation and use this concept both for descriptive and theoretical purposes.

The concept of national systems of innovation is, however, only relevant in those cases where you can identify clear national differences in institutions and economic structures. This is not always the case. Innovation may be systemic without being national and sometimes it may be more fruitful to analyze regional and local systems of innovation. Sometimes sectorial delimitations, as for instance 'technological systems' (Carlsson, this volume) are more useful. The different concepts and approaches might be said to complement more than substitute for each other and it is often the specific empirical context that dictates which one to choose.

In a way the national system of innovation approach to innovation theory and innovation studies is inherently comparative: There is no way in which an optimal system of innovation can be defined since the process of innovation always to some extent is a venture into an unknown future characterized by genuine uncertainty. Furthermore, it is an open-ended process without a final equilibrium position. It also follows from the definition of national systems of innovation that they must be specific to time and space. They must have a local history, and can only be evaluated in relation to this.

There is, thus, no equivalent here to the concept of 'efficient allocation of resources' in the usual static framework. The best one can do is to compare different ways in which structures of production and institutions affect innovation in different periods and countries (regions, locations) and try to learn something about the processes of technical and institutional change from this.

6.1 Methodological Considerations

6.1.1 How Can National Systems of Innovation Be Compared? - A General Comparative Model

It is certainly not self-evident, however, how such comparisons could, or should, be made (Mjøset 1992). It is not obvious what characteristics of the national system of innovation one should concentrate on or how they should be measured. Comparisons between different national systems of innovation are important elements in the further development of the concept and necessary for it to have any policy relevance, and it is important to be careful about methodological problems in this connection.

It follows from the character of innovation, and from our definition of national systems of innovation, that we should concentrate on, and at the same time make a distinction between, structural and institutional factors when comparing the determinants of innovation in different countries. It is also useful to separate between the determinants of innovation, on the one hand, and innovation as an outcome or performance factor, on the other. Identifying a performance aspect of the national system of innovation also makes it relevant to identify policy instruments as separate innovation determinants. Finally, the historical character of innovation processes must be recognized, which means that the dynamic interdependence between the factors affecting innovation and the time paths of both these factors and the factors describing the performance of the system should be taken into account.

Thus, to be able to describe and compare national systems of innovation we believe that we need the following variables:

'O' symbolizes the 'outcome' or 'performance' of the national system of innovation. It can be thought of as a list, (o1, o2,...,on), of several elements, all dated and measured over a relevant period. In the list of performance variables we include factors to which we attach some positive or negative value. We are not interested in national systems of innovation for their own sake, but because they relate to aspects of the economic process, which we include in some type of ultimate 'preference function'. Through the performance factors the normative aspects of studies of national systems of innovation are brought into attention and the value premises are, partly, made explicit.

'I', (i1, i2,...,in), is a list of the most important factors in the institutional set-up which, according to our definition, shape the communication and interaction between people in and between firms and other organizations and thus the interactive learning processes from which innovations result. The institutional set-up should also be dated and measured over a historical time.

'S', (s1, s2,...,sn), is a list, properly dated and measured over time, of factors in the production structure which, together with the institutional set-up, constitute the physical base and knowledge base from which innovations proceed in trajectories as cumulative processes.

'P', (p1, p2,...,pn), is a dated list and description of the use of, (local regional and national) government policy instruments that together with I and S, affect the innovation process.

'E', (e1, e2,...,en) represents a group of exogenous factors which influence the innovation process, but which are beyond the direct control of the government or other domestic actors. We can think of the national system of innovation as surrounded by an environment and we can call the factors constituting this environment, simply, exogenous factors.

Against this background we propose that comparisons between different national systems of innovation can, fruitfully, be done in the following way.

Let (I_t, S_t, P_t) represent a 'complete' description of the innovation properties of an economy in a given environment, E_t, at a certain point in time, t. The change over time of these innovation properties can then be formulated as $d(I, S, P)/dt = g(I_t, S_t, P_t, E_t)$;

If we know 'g', we can also formulate a performance function $O_t = f(I_t, S_t, P_t, E_t)$ which together with $d(I, S, P)/dt = g(I_t, S_t, P_t, E_t)$ describes the essence of the national innovation system, i.e. how structural and institutional factors in a process over time affect innovation. They relate the history of the performance of the innovation system to the history of its institutional and structural components and to the innovation policies pursued.

The relationship $d(I, S, P)/dt = g(I_t, S_t, P_t, E_t)$ is, of course, complicated and non-linear. Its precise form can only be considered stable for limited periods. It can change, sometimes in jumps, and it is different between countries. It is a dynamic relationship and it must allow for cumulativeness and path-dependency as well as unpredictability and accidental combinations of knowledge in the innovation process. It might be thought of as describing some kind of evolutionary process, where evolution is both of the Lamarckian 'learning' kind and of the totally blind, small step groping Neo-Darwinian kind and at the same time allowing for 'macro-mutations' in the form of radical innovations.

Even if we acknowledge these difficulties and admit that we know very little about the concrete forms in specific countries of the relationships above (i.e. $O_t = f(I_t, S_t, P_t, E_t)$; and $d(I, S, P)/dt = g(I_t, S_t, P_t, E_t)$), they can be of considerable help in comparisons between national systems of innovation as a frame of reference, as a structuring and focusing device and as a check-list. In this way we emphasize the dynamic and historical character of national systems of innovation.

If we want to describe and compare how institutions, economic structures and innovation-related policies determine innovation performance over time in different countries, we can refer to this as 'complete, or over-all, comparisons', i.e. the comparison involves the whole relationship and includes elements of institutions, structures, policies, exogenous factors and performance.

If we want, for example, to limit the comparison to some performance elements or some policy elements, or maybe both, we can refer to this as 'partial comparisons'. We should then be in a position to see that efforts to explain performance from policies alone, i.e. from an '$O = f(P)$' relationship, are incomplete and may be misleading. In this way the general comparative model help to structure comparisons, so that we are clear about what we do and do not compare, and abstain from premature conclusions about where specific national systems of innovation have their strengths and weaknesses.

6.1.2 Some Specifications of the Model

A few comments and examples may serve to make the model look a bit less abstract and general.

The performance of a national system of innovation can be divided into 'immediate performance' and 'ultimate performance,'[1] 'Immediate performance' refers to

[1] By these terms we mean that there is a causal rather than a time relationship between immediate and ultimate performance.

measures of direct results of the innovation system in terms of product- and process innovations and their diffusion. Since good direct measures of innovations are hard to come by, we usually have to do with indications as for example patents, new products as a share of total sales and the proportion of high-tech products in production and foreign trade.

'Ultimate performance' refers to the contribution of innovation processes to the basic variables in the 'social welfare function' as for instance levels and growth of income, employment and the balance of payments. It can also refer to more specific performance indicators as environmental standards, working conditions, social security, equity, etc. In order to make a distinction between the performance of an economy in general terms and the performance of the national innovation system, we might limit the latter to relate to the dynamic efficiency and international competitiveness of the economy and, thus, ultimately, to the innovation process. The ability of an economy to grow under high employment and without accumulating foreign debt while observing strict environmental standards might be taken as an example of a measure of 'ultimate performance' of the innovation system. This still leaves us with many problems, however, and for both theoretical reasons and from an innovation policy point of view, there is a need to develop better indicators of both ultimate and immediate innovation performance.

The very long list of factors in the institutional set-up which affect innovations, can be divided into 'formal' and 'informal' institutions. 'Formal institutions' refers to formally organized or codified entities. The formal system of education and training, the formal research system, the telecommunication infrastructure, the system for financing innovations, the technical service system, the patent-, trademark and copyright systems and other aspects of the system of property rights are some obvious examples. Post offices, labour unions, government agencies and other tangibles which are generally referred to as institutions in everyday speech may also be referred to as 'formal institutions' under this definition.[2]

'Informal institutions' refer to those habits, routines, rules, norms, etc. which affect the character and pattern of communication and thus interactive learning. The degree of working place democracy, the degree of trust and opportunism and the informal cooperation norms in both intra-firm and inter-firm relations, informal network relationships and user-producer relationships, etc. can be mentioned as examples.

Both formal and informal parts of the institutional set-up are important elements in the national systems of innovation. The problem is, of course, that they are difficult to measure in ways which make comparisons possible. In addition, for many informal institutions there are no documented or reliable observations at all.

[2] To say that a bank is a financial institution actually means that already institutionalised acts of borrowing and lending (i.e. borrowing and lending according to certain rules, keeping reserves, respecting certain norms for interest and debt payments and so on) have been formalised and organised by the creation of a bank.

When comparing the 'production structure' as a determinant of innovation, both the industrial production structure in general and the export specialization structure on different aggregation levels are relevant. Also other factors like firm size distribution, degrees of concentration and centralization, and size and structure of public investments and consumption might be considered.

The technological level of the economy, embodied in both its capital stock, its work force and its organizational routines should also be included. This can be done in at least two ways. One way is to measure the relative weights of high-, medium- and low-technology sectors in production and trade. This gives a picture of how well the production structure has adapted to technological upgrading and demand trends in the global economy. Another way is to make a cross-sectional description of technological levels for the main parts of the economy. There are different ways to do this. One can use a 'Delphi method', i.e. ask people who know a lot about the state of technological art in different industries, or one can measure the diffusion of special advanced key technologies, for example microelectronics-based production technologies, through the economy.

We may also include the economy's resource base in a description of its structure. Obviously the capital stock in a broad sense, i.e. including natural capital and human capital as well as buildings, production equipment, etc. influences innovation processes and channels it into certain directions. This is one reason for the cumulative character of innovation.

Alternatively, we may treat the resource base as a specific factor in the over-all comparative model. This may be convenient for countries (like Finland, Sweden and Denmark) in which natural resource-based production complexes tend to have lasting influences on not only production but also on innovation structures. Since it is probably not possible to measure the resource base directly in terms of stock variables (for example wealth) it will probably have to be indicated by measures of the structure of resource-based production.

Since most countries have some policies, often targeted at high-tech areas, directly aiming at stimulating innovation, it is relevant to include this in the comparison of different national systems of innovation. However, other types of policies, which are not directly targeted at innovation activities, affect the innovation system even more. Education policies, income distribution policies, social security policies, employment policies, policies in relation to the communication infrastructure, etc. all affect interactive learning and innovation. Comparisons of national systems of innovation have to reflect this. The borderlines between descriptions of such policies and of the institutional framework may be rather fuzzy, but the introduction of a policy-variable in the general comparative model illustrates the idea that it is, normally, easier to change policy under a given institutional set-up than it is to change this set-up.

One very important set of phenomena which is not explicitly included in the general comparative model but has to do with the dynamic relations between its variables and which very much affects the performance of a national system of innovation is the degree of institutional-technical match (or mis-match). This term refers to how well the economic structure and its technical characteristics (the present technological level of its equipment), fit with the patterns and styles of communication and interactive learning in the economy and with the organizational forms of firms.

The match/mis-match can be described in terms of how well the institutions and organizational forms adapt to and utilize technological change. These matters are central to the understanding of the dynamics of national systems of innovation. Especially in periods of radical technical change they become crucial for the performance. For example, different countries cope with the transition from fordist to post-fordist growth patterns very differently.

This may have something to do with the diversity of the innovation system; its institutional diversity and the diversity of its production structure (Johnson 1992). A 'learning perspective' on innovation is closely related to an evolutionary perspective, and it is a basic proposition in evolutionary theory that the diversity of a system affects its development.

Generally speaking, diversity affects innovation because it affects technical, organizational and institutional learning and contributes to the knowledge base of the economy. Diminishing this diversity means destroying parts of the economy's stock of knowledge and reducing the number of technical options. It also means decreased possibilities for communication and interaction between different kinds of skills, knowledge, and competence and, thus, reduced learning possibilities. Diversity generates novelty and affects the learning capability of the economy. It also affects the ability to cope with exogenous disturbances and threats to the innovation systems. Fluctuations in world markets, innovations, political movements, ecological stress and so on are often quite unpredictable. No system can cope with each and every possible kind of disturbance, but systems with a lower degree of diversity or rather, with a lower capacity to generate new diversity, are more vulnerable. Flexibility depends on a diversity generating mechanism, and generation of diversity is a way to handle uncertainty and works as a shock absorber. Against this background, diversity in the national systems of innovation may be a condition for a successful transfer to a new techno-economic paradigm. It should be observed, however, that there is also a cost of diversity for example in terms of lost economies of scale in both production and innovation. The need for concentrated efforts in many research areas puts an effective limit on diversity in the national system of innovation, especially for a small country, which will be discussed in the next section.

The discussion so far has shown that, from a methodological point of view, comparison between different national systems of innovation is an endeavour which vastly exceeds simple counting of patents, R&D expenditures, and the number of engineers and scientists. It is very difficult to make short and partial comparisons of such systems without risking erroneous or biased conclusions. However, we are not able to do much more than this. We still need much conceptual clarification and there is an extreme lack of both quantitative and qualitative data. It is, therefore, with a somewhat uncomfortable feeling that we now proceed to the more empirical part of this paper. It must be kept in mind that we are not able to do anything more than a rather sketchy and very selective comparison of the basic variables in our comparative model.

6.2 Comparing Denmark, Finland and Sweden, Some Points of Departure

As a first step in a comparison between the national systems of innovation of Denmark, Finland and Sweden it may be helpful to discuss some obvious similarities and differences, which can serve as a point of departure.

6.2.1 The Environment of the National System of Innovation

The geo-political and trade-political position of the country are obvious examples of exogenous factors where there are important differences between Denmark, Finland and Sweden in spite of their close geographical proximity. In connection with defense- and foreign policies there have been sharp borderlines between the Nordic countries in the whole period after 1945, and partly as a consequence of this, trade relations with the Soviet Union have been much more important for Finland than for the other countries.

Other examples of exogenous factors are the main elements in the development of the world economy. This is true especially for relatively small and open countries. Being small, open and industrialized is of course an important similarity between Denmark, Finland and Sweden, which affect their innovation systems in several ways. It can be argued that it makes a diversified and broad utilization of the human factor in innovation more important, since it is relatively difficult for a small country to rely on science-intensive innovation processes because of economies of scale and scope in research and development. Broadly based interactive learning processes leading to incremental innovations over a wide spectrum of products and process are relatively important parts of a small country's innovation system.

For the same reasons small countries have to be more specialized, when it comes to R&D-based innovations. It is not possible for them to commit R&D resources to many different areas, but it might be possible, and maybe necessary, to apply interactive learning and innovation as a normal 'mode of operation' in most parts of the economy. This may also be a way to reduce their vulnerability to structural changes in the international division of labour.

The institutional systems of small homogeneous countries have some common characteristics. It is likely to be relatively transparent. The 'rules of the game' are well known and there is a rather small number of important decision-makers in an environment of relatively low communication costs. These factors may make institutional change, including 'institutional borrowing' from other countries easier, even if it also creates a risk for a kind of institutional stalemate where different interests block each other and prevent institutional change. The need for an institutional system which stimulates interactive learning and innovation is relatively strong for small countries. The possible benefits of such a system are considerable, and so are the potential costs of institutional rigidity. This brings the communitarian elements of the institutional systems of the Nordic countries into focus.

6.2.2 Common Institutional Preconditions for Interactive Learning

It can be argued that the Nordic countries share some characteristics which might help interactive learning: They have efficient communication infrastructures; they are rich countries with moderate income inequalities; they are, relatively, culturally homogeneous with solid democratic traditions; they have well developed systems of education; they have efficient systems of technological service; the social security systems and health care systems are of a comparably high quality; consumer norms are similar; the labour markets are organized and with well developed institutions for consensus seeking and handling of conflicts (even if Finland earlier has had a significantly lower level of class consensus than Denmark and especially Sweden) and considerable amounts of money are spent on labour market policies, and so on (Edquist and Lundvall 1993).

It is relatively easy to go on like this and list common characteristics, which, at least for a distant observer, may make it look as if the Nordic countries have rather similar institutional preconditions for interactive learning and innovation. Still, a closer look will probably reveal important differences in innovation systems and it is, of course, one of the purposes of comparative analysis to do just this.

6.2.3 Common Structural Preconditions

The same is true for the structural determinants of the innovation process. All three countries were late starters but fast runners within the group of present industrialized high-income countries. Roughly speaking the aggregate (agriculture-industry-service) production structure is now similar[3] among the three countries. It is important to observe, however, that the rate of structural change has been extremely quick in the last decade.

The industrial development in all three countries has to a significant degree been resource based: In both Finland and Sweden a large part of industrial exports has

[3] According to Maddison (1991) The percentage A-I-S structure of employment in 1987 were in Finland 10-31-59; in Denmark 6-28-66; and in Sweden 4-30-66.

had its origin in forestry and mining. The export of more or less refined raw materials has continued to play a role and production and export of machinery related to natural resources have become main positions of strength in the industrial structure. For Denmark we see the same picture in relation to the agricultural sector.

It is of course an important question whether this will turn out to be a weakness for the national systems of innovation in an information technology-based new techno-economic paradigm. It should be observed that in a period of radical technical and institutional change national systems of innovation will come under strain and will have to adapt more or less drastically. If the need for flexible institutional systems is relatively strong for small countries, this need will be even more strongly felt for 'small countries facing the technological revolution' (Freeman and Lundvall 1988). There is no doubt that such changes are both fast and numerous in the present period in Denmark, Finland and Sweden (in the system of education, in the welfare state institutions, in the organization of markets, in the structure of ownership, etc.), but it is also extremely difficult to know how well the Nordic national systems of innovation will adapt and cope with the shift in the techno-economic paradigm. Both the need for comparative studies of national systems of innovation and the methodological difficulties in carrying them out increase.

Following the logic of our O = f(I, S, P, E) relationship we now turn to presenting brief (and highly selective and rather unfinished) descriptions and comparisons of institutional, structural and policy factors for Denmark, Sweden and Finland. (For the external factors we have to refer to the brief discussion in section 6.2.1 above.) After this the performance of the three national systems of innovation will be compared.

6.3 Comparison of Economic Structure

The purpose of this section is to locate some country-specific structural characteristics that may turn out to be 'critical' in shaping and maintaining long-term innovative capabilities.

The overall economic production structure is normally described in terms of distribution of employment or production by industries (for instance Agriculture, Manufacturing, Public and Private Services). Using this approach we see, as stated in the previous section, that the production structures in the three countries have converged and now appear rather similar.

Recent studies of specialization and international competitiveness stress the importance of interdependencies between firms and industries, reflecting that obtained international strongholds often appear in clusters of related industries and firms rather than as isolated islands (Brændgaard et al. 1984, Andersen and Lundvall 1988, Porter 1990, Dalum et al. 1991). In our general description and comparison of the economic structure we will use such a 'systemic' approach and focus on a few dominating development blocks within the three countries.

By a *development block* we mean 'a broad set of interconnected producers and users of products, developing in close interaction with each other and often supported by knowledge-producing private, or public, organizations' (Edquist and Lundvall 1993). We define development blocks at the national level, i.e. they consist of *domestic* industries coupled by strong quantitative and qualitative linkages, be they vertical (user-producer relations) or horizontal (within the same industry).[4]

Development blocks evolve over long periods of time and they seem to be rather 'tenacious of life', not least because formal and informal institutions and various interest groups surround and often embalm them.

The characteristics of the present pattern of specialization in the three countries have, as we shall see in the following, strong and deep roots in the history of the transformation of a few dominating development blocks. *Firstly*, we stress the relatively mature resource based development blocks of agro-food, forestry-paper and mining-metal. *Secondly*, we emphasize the rise of the engineering block in the three countries. *Thirdly*, we focus on the development of two (still) relatively small, but fast growing development blocks, telecommunication and health.

However, it is difficult to identify such development blocks in quantitative terms. The relevant borders surrounding a development block are often blurred and difficult to identify unambiguously. Furthermore, we lack statistics to support this approach analytically.[5] Our short description and comparison are thus mainly based on a rather intuitive or heuristic classification of industrial development blocks or clusters of interrelated commodity flows.[6] We use international trade-statistics (export structure and export specialization) as rough indicators of some of the strongholds and weaknesses in the present innovation systems. One reason why we focus on export specialization[7] when we describe the economic structure

[4] The term 'development block' was originally introduced by Dahmén in the 1950s. For a discussion of development blocks and their structural tensions in relation to national systems of innovation see Andersen (1992).

[5] Input-output tables give some limited possibilities to identify development blocks based on inter-industry flows of commodities.

[6] The classification of the development blocks and the corresponding trade-statistics used in this chapter is based on calculations on the IKE trade database at the Institute of Production, Aalborg University and draw heavily on resent work by Bent Dalum. Bent Dalum, 'The National Framework' in Rob van Tulder (ed.), *The Competitive Advantage of Welfare States - Small Countries' Ways out of the International Restructuring Race*, forthcoming.

[7] Export specialisation figures (The Balassa Revealed Comparative Advantage Index) is defined as:

$$\frac{Xij / Xj}{Xi / X}$$

where Xij are exports from country j of commodity i; Xj are total exports from country j; Xi are total OECD exports of commodity i; and X are total OECD exports (Dalum 1992).

is that economies become increasingly open. Nevertheless, patterns of specialization are distinctively different from country to country, and they seem to show a considerable stability over time. On the other hand, the learning and innovation effects of different specialization patterns are important determinants for differences in the overall competitiveness of nations (Andersen 1992).

6.3.1 Resource-Based Development Blocks

It is remarkable how the *resource-based* industrial development blocks have dominated the production structure throughout the whole industrial transformation period since the late nineteenth century in the three countries, although the importance is less manifest in Sweden than in Finland and Denmark.

The Danish *agro-food* block has developed and maintained strong domestic linkages through nearly every part of the value chain from grain and milk to butter cookies and fancy fast-food menu's, i.e. both the foodstuff and a wide range of the machinery, the chemicals and the R&D institutions necessary to produce and develop agro-food related products. Though the share of the agro-food block of the total Danish employment, production and exports strongly decreased during the sixties, around one third of the Danish exports still come from this block at the end of the eighties, see table 6.1.

Table 6.1. Export structure Denmark. Per cent of total exports

Development block	1965	1969	1973	1979	1984	1987
Agro-food	54.74	43.20	42.19	38.75	36.09	34.54
Forestry-paper	4.30	4.68	4.75	5.34	7.23	6.69
Textile-clothing	4.57	6.35	6.11	5.21	5.39	5.92
Metal	3.68	4.61	4.81	6.47	5.88	5.73
Health	1.91	2.21	2.46	2.83	3.83	4.61
Telecommunication	0.87	1.24	1.36	1.33	1.22	1.69
Rest: resource based	5.53	6.22	5.83	6.71	5.20	5.44
Oil & gas	0.91	1.18	1.93	3.68	4.86	2.80
Rest: chemicals	3.45	5.92	4.26	5.32	6.33	6.12
Rest: engineering	16.94	20.57	21.80	19.98	19.34	20.20
N.e.s.	3.10	3.82	4.49	4.38	4.62	6.27
Total	100.00	100.00	100.00	100.00	100.00	100.00

Source: IKE trade database/Bent Dalum

Table 6.2. Export specialization Denmark. Balassa Index

Development block	1965	1969	1973	1979	1984	1987
Agro-food	3.23	3.18	2.68	2.82	2.86	3.09
Forestry-paper	0.67	0.77	0.85	1.03	1.48	1.23
Textile-clothing	0.47	0.70	0.69	0.78	0.92	0.91
Metal	0.24	0.31	0.35	0.50	0.57	0.61
Health	1.46	1.51	1.69	1.81	2.21	2.15
Telecommunication	0.78	1.04	1.12	1.05	0.82	0.99
Rest: resource based	0.86	1.04	1.01	1.06	0.95	1.04
Oil & gas	0.40	0.54	0.71	0.76	0.72	0.78
Rest: chemicals	0.46	0.75	0.55	0.55	0.68	0.64
Rest: engineering	0.61	0.63	0.67	0.61	0.54	0.52
N.e.s.	0.60	0.73	0.97	0.85	0.86	0.97
Total	1.00	1.00	1.00	1.00	1.00	1.00

Source: IKE trade database/Bent Dalum

The export specialization figures in table 6.2 confirm the present importance of this block for the Danish economy. If we go behind these relatively aggregated figures, we will, of course, find great differences in the degree of export specialization among the various product groups in the block, but no single product group with a specialization figure below 1.00 in 1987. Despite the success of the Danish agro-food block its existence also reflects serious structural problems for the economy due to falling or stagnating demand for agro-food products. When, in 1972, Denmark joined the EEC, the most traditional elements in this block were reinforced and a potential reconstruction was delayed. In recent years a series of mergers has taken place within the block, but it is still an open question whether this kind of regrouping will give the necessary renewal and impetus to a more offensive and innovative strategy based on, for instance, biotechnological applications in the agro-food area (Edquist and Lundvall 1993).

Another important resource-based development block in the post-war period relates to *domestic construction*. Despite serious building crises in the seventies and eighties, we still find some Danish technological strongholds left in cement, cement processing machinery and production systems and various other building materials. (In table 6.2 these strongholds are contained in both the 'rest: resource-based' product group and the heterogeneous 'rest: engineering' group.)

From table 6.3 to 6.6 it is clear that an export-oriented development block has emerged around *forestry, pulp, paper and related machinery* in Sweden and Finland. The share of the total Swedish exports from this block has been falling during the post-war period and amounts to around one fifth in 1987, but its present importance and international competitiveness are still reflected in the highest export specialisation figures (4.01) in the country, see table 6.6. Although the share of this block of the total Finnish export has declined dramatically from 70% in the middle of the sixties to around 40% in the late eighties (cf. table 6.3) it is still within this block that we find extremely high export specialization figures (cf. table 6.4). In an international comparison the forest products industries in Finland are doing rather well. Today Finland is the largest exporter of printing and writing papers in the world.

Table 6.3. Export structure Finland. Per cent of total exports

Development block	1965	1969	1973	1979	1984	1987
Agro-food	5.36	5.33	4.80	4.07	4.93	3.89
Forestry-paper	70.33	60.85	53.54	45.44	38.88	41.53
Textile-clothing	2.89	5.91	8.02	6.84	5.48	4.90
Metal	6.71	7.76	8.81	10.86	9.33	9.84
Health	0.06	0.16	0.26	0.47	0.71	0.92
Telecommunication	0.17	0.20	0.56	0.37	0.87	2.03
Rest: resource based	2.80	2.98	3.09	4.37	3.57	4.13
Oil & gas	0.01	0.42	0.17	2.35	5.43	2.15
Rest: chemicals	1.61	2.24	2.92	4.44	5.12	5.09
Rest: engineering	8.98	12.02	14.97	17.33	22.34	21.98
N.e.s.	1.07	2.12	2.85	3.45	3.34	3.53
Total	100.00	100.00	100.00	100.00	100.00	100.00

Source: IKE trade database/Bent Dalum

Table 6.4. Export specialization Finland. Balassa Index

Development block	1965	1969	1973	1979	1984	1987
Agro-food	0.32	0.39	0.30	0.30	0.39	0.35
Forestry-paper	11.03	10.40	9.60	8.76	7.96	7.66
Textile-clothing	0.30	0.65	0.91	1.03	0.94	0.75
Metal	0.44	0.53	0.64	0.85	0.90	1.05
Health	0.05	0.11	0.18	0.30	0.41	0.43
Telecommunication	0.15	0.17	0.46	0.29	0.59	1.19
Rest: resource based	0.43	0.50	0.53	0.69	0.65	0.79
Oil & gas	0.01	0.19	0.06	0.49	0.81	0.60
Rest: chemicals	0.21	0.29	0.38	0.46	0.55	0.53
Rest: engineering	0.32	0.37	0.46	0.53	0.62	0.57
N.e.s.	0.21	0.41	0.62	0.67	0.62	0.55
Total	1.00	1.00	1.00	1.00	1.00	1.00

Source: IKE trade database/Bent Dalum

Table 6.5. Export structure Sweden. Per cent of total exports

Development block	1965	1969	1973	1979	1984	1987
Agro-food	5.96	5.22	4.90	4.31	4.28	3.55
Forestry-paper	29.57	25.91	25.65	22.21	21.44	21.74
Textile-clothing	3.05	3.55	3.51	2.87	2.51	2.65
Metal	20.85	21.05	18.13	16.87	14.52	13.04
Health	0.74	0.96	1.14	1.75	2.16	2.66
Telecommunication	2.01	2.71	3.20	3.32	3.46	3.16
Rest: resource based	3.25	3.19	3.13	3.12	2.80	2.86
Oil & gas	0.41	0.83	0.53	2.64	5.30	2.62
Rest: chemicals	3.12	3.61	3.58	4.68	5.23	5.51
Rest: engineering	28.79	30.63	34.05	35.02	35.46	38.41
N.e.s.	2.26	2.33	2.17	3.22	2.84	3.79
Total	100.00	100.00	100.00	100.00	100.00	100.00

Source: IKE trade database/Bent Dalum

Table 6.6. Export specialization Sweden. Balassa Index

Development block	1965	1969	1973	1979	1984	1987
Agro-food	0.35	0.38	0.31	0.31	0.34	0.32
Forestry-paper	4.64	4.28	4.60	4.28	4.39	4.01
Textile-clothing	0.31	0.39	0.40	0.43	0.43	0.41
Metal	1.37	1.44	1.31	1.32	1.40	1.39
Health	0.56	0.65	0.78	1.12	1.25	1.24
Telecommunication	1.79	2.28	2.64	2.64	2.33	1.85
Rest: resource based	0.50	0.54	0.54	0.49	0.51	0.54
Oil & gas	0.18	0.38	0.20	0.54	0.79	0.73
Rest: chemicals	0.42	0.46	0.46	0.49	0.56	0.58
Rest: engineering	1.03	0.93	1.05	1.07	0.98	0.99
N.e.s.	0.44	0.45	0.47	0.63	0.53	0.59
Total	1.00	1.00	1.00	1.00	1.00	1.00

Source: IKE trade database/Bent Dalum

The international competitiveness of the firms in the Finnish forestry-paper block reflects the results of a long historical development process where Finnish firms gradually have built up their innovative capability from a basis where most machinery and process technology were imported (1850-1914) to a phase with a high degree of import substitution of and national self-reliance on strategic machinery and process technologies (1918-1957) to today's position among the world export leaders and technological avangardists (Raumolin 1992). Similar to the development of the Danish agro-food block, the innovative maturity process of the Finnish forestry-paper block is a result of several interdependent structural, institutional and political factors internal and external to the Finnish economy: Interactive learning between producers of pulp and paper, suppliers of machinery and process control equipment, and linkages between the mining sector and the pulp and paper industry (sulphur); an active role of the central government and state-owned companies; extensive private and public R&D facilities and education of engineers; a building up of necessary infrastructure to mention some of the important factors behind the success (see Raumolin 1992).

A second relatively large and mature resource-based development block in both Sweden and Finland is the *metal* block. Similarly to the forestry-paper block domestic linkages between the producers of raw materials (e.g. iron and steel) and the suppliers of mining equipment have historically supported interactive learning and innovation (Edquist and Lundvall 1993, Raumolin 1992). In Sweden this block has declined from around 20% of the total Swedish exports in the sixties to 13% in 1987. In Finland, on the contrary, the metal block has increased its export share in the period from 7% to approximately 10%.

The relatively low growth potential within the mature resource-based development blocks is an important part of an explanation of the present structural problems in the three countries. As in the case of Denmark, these development blocks may also in Sweden and Finland bear the seeds of severe structural problems if they fail to reconstruct old and develop new strongholds. The prospective demand for the dominating resource-based products (wood, pulp, paper, iron ore, steel, etc.) may stagnate because of the rapid development in biotechnology and material technology. Mastering these new technologies and finding a solution to the growing environmental problems are important challenges for a strategic renewal of the resource-based development blocks in all three countries.

6.3.2 'Engineering'

Various engineering products form a growing part of the total exports in each of the three countries, as in most other OECD countries. The low degree of export specialization in engineering in Denmark, Sweden and Finland (table 6.2, 6.4 and 6.6) does reflect the specific definition of this group, where various machines for agriculture and food processing, paper and pulp machinery, and metal working machinery are included, respectively, in the agro-food block, the forestry-paper block and the mining-metal block. It is especially within these areas of the engineering sector that international strongholds exist. In Finland, for instance, more than one fifth of the total exports of the metal and engineering industry consists of forest industry-related products (Vuori and Ylä-Anttila 1992).

The remaining groups of various engineering products in table 6.1 to 6.6 are very heterogeneous and do not form proper development blocks as defined above. For further analysis it would be relevant to divide this large group into sub-blocks using a 'technological system approach' as introduced by Carlsson in his chapter in this volume. As suggested in Edquist and Lundvall (1993), in Sweden such a division of the engineering industry might result in a number of development blocks centered around the large Swedish engineering firms or parts of them, for instance an electricity block, an automobile block and an airplane block. In Denmark and Finland, however, the engineering group in table 6.1 to 6.6 mainly consists of relatively small firms producing traditional and less R&D-intensive machinery or components for engineering products of larger domestic or foreign firms.

6.3.3 'New Blocks'

There is without doubt a strong need for renewal of the dominating resource-based development blocks and a need for new strategic blocks. Next to the forestry-paper block the *telecommunication* block in Sweden has the highest export specialization figures throughout the post-war period reflecting international strongholds based especially on regulation, standardization and public procurement policies (Grandstrand and Sigurdson 1985). The Finnish 'electronic adventure' in the form of rapid diffusion and application of basic technology to selected areas within the old production system (for instance process control equipment for the pulp and paper industry) may, at first, look as an exemplary case of a successful renewal process (Lemola and Lovio 1988). However, according to Vuori and Vuorinen (see Introductory Chapter) the growth has already reached its peak, and some of the most advanced units have either been sold to multinationals or been closed down. In an international comparison, the foreign share in the Finnish electronics industry (measured in terms of employment) is, however, still relatively small (Lovio 1992). Lovio concludes that the overall influence of foreign electronic companies has been positive for the Finnish economy. Up to the middle of the 1980s, foreign electronic companies served as an important channel for the diffusion of new technology in Finland. By now this technology import role has changed to a technology exporting role.

The expanding *health* block (consisting of medical industry, medico electronics and other types of medical equipment) is an example of a relatively new and interesting growth area in the Danish system of innovation and to a lesser degree also in Sweden, while it (still) seems negligible at least in quantitative terms in Finland. The international strongholds of the Danish medico industry are to a high degree rooted in links to an advanced domestic hospital and health care system (Jørgensen 1986, Andreasen and Lotz 1990). The Danish hearing-aid industry, for instance, obtained its international strongholds (today about 2/3 of the world market) by a combination of high level technical standards within the electro-acoustics area with a solid domestic market based on public consumption (Jørgensen 1986). Similarly, a qualified public sector demand and a relatively strict regulation have played a significant role for the innovative capability of the Danish environment industry and its international strongholds (Gregersen 1992). It is obvious that the role of the welfare state as a competent user has been crucial in the development of these new blocks, but we need more detailed comparative studies before policy conclusions may crystallize.

This very brief description of the dominating development blocks indicates that the three countries may face fundamental structural problems because a substantial part of their production is specialized in products with falling or stagnating demand, unless a high income elasticity-oriented diversification process in relation to the blocks can be created. The germinating new blocks are all still too small to compensate for the shrinking old colossuses.

Another indicator of the need for a renewal process in the three countries may be the proportion of high-tech products in foreign trade. Often this indicator is used as a performance measure, but here we apply it as a structural characteristic, e.g. as a measure of the technological level. High-tech is defined in terms of R&D intensity within specific industries and product groups.

The proportion of new high-tech products in total exports is especially low in Denmark (9.0%) and Finland (10.8%), even though Finland has been able to improve her share considerably in the last decade (cf. table 1.13 in chapter 1). Sweden is placed in the middle group together with Austria, Germany and France. With the exception of Japan, which has a comparatively low high-tech import share, the import figures for the other countries are remarkably similar.

Since R&D-intensive sectors are characterized by higher demand growth, a higher productivity level and a higher productivity growth than manufacturing as a whole (Edquist and McKelvey 1991) the relatively low high-tech proportion in the three countries may be an indicator of long-term structural problems.

6.3.4 Firm Size

Another structural factor which influences the innovative capability, especially in scale-intensive R&D areas, is the firm size. Out of Europe's 500 biggest manufacturing firms in 1988 only 4 are located in Denmark, 14 in Finland, but 33 in Sweden. If the number of large companies is related to the size of the economy (population), however, both Sweden and Finland are ranked high; respectively third and second in the OECD area (ETLA, IFF, IUI, NØI 1990: 56). The Danish firm structure is clearly dominated by many small and medium-sized firms with low R&D intensity (cf. section 6.6.1), and this may hamper the long-term innovative capability of the innovation system. In contrast, the Swedish firm structure is characterized by relatively large firms with a high R&D intensity, which ceteris paribus should stimulate innovation activities.[8] The average size of the top 8

[8] Average plant size in manufacturing 1986 (number of persons engaged): DK 57; SF 75; S 85. Average plant size in engineering 1986 (number of persons engaged): DK 64; SF 80; S 99 (ETLA, IFF,IUI & NØI 1990:34)

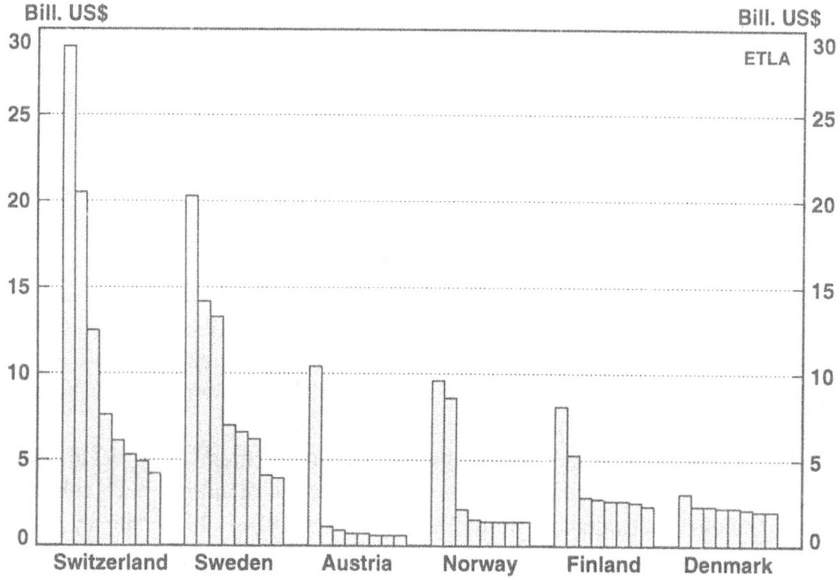

Fig. 6.1. Individual sales of the 8 largest manufacturing companies in selected countries, 1989. Source: Heum and Ylä-Anttila (1992), p. 37.

manufacturing corporations in Denmark and Finland is still less than half of that of the Swedish companies, see figure 6.1.

During the 1980's the share of large companies has grown in the three countries as a consequence of takeovers and mergers, often motivated by the ongoing European integration process.

Firm size may be important in relation to innovative capability for several reasons. Firstly, the importance of economies of scale in R&D and innovation varies across industries and technology areas, which means that cross-country comparisons of R&D intensity and firm sizes should take into account differences in industrial structure.

Secondly, large firms seem to play an important role both as 'growth engines' and as spurs in the process of internationalization. Heum and Ylä-Anttila (1992) found that the growth of the 30 largest industrial corporations in Denmark, Finland, Norway and Sweden over the period 1974-1990 was significantly faster than that of the total manufacturing. Although the main part of the employment growth has been outside the national borders the contribution of the largest companies to total domestic manufacturing employment has increased slightly.

The Swedish industrial companies are still the far most internationalized, but the rate of internationalization of large industrial companies has been particularly fast in Finland (and Norway) during the 1980's (Heum and Ylä-Anttila 1992). On the one hand, this growing internationalization may stimulate technology transfer across boarders as mentioned above in the Finnish case. On the other hand, it may also hamper interactive learning and the innovation processes in the long run if established domestic linkages are broken, for instance if domestic suppliers are replaced. We clearly need longitudinal studies of how user-producer relations and other linkages develop and effect the innovative capability of both large and small firms regarding increasing internationalization.

6.4 Comparison of Institutional Set-up

As discussed in section 1 there are a variety of factors which, ideally, should be included in a comparison of the institutional set-ups of different countries. However, since it is not possible to make a complete comparison of the institutional set-up, we again have to select a few factors which we believe are of special importance.

The following sub-section deals with formal institutions and we include the formal research system, the education system, the vocational training system, relationships between firms and universities, the system of technological service and the financial system. Section 6.4.2 discusses the role of informal institutions.

6.4.1 Formal Institutions

The R&D System. One of the central formal institutions influencing innovation is how, and to what extent, a country organizes the generation of new knowledge through research and development. Therefore we shall start with a discussion of the 'research systems' of the three Nordic countries.

In all Western countries R&D is performed by both private firms and public institutions. In Finland, Denmark and Sweden the set-ups of the institutions carrying out R&D are rather similar and therefore R&D can be further sub-divided into R&D carried out by industry and other private sectors, R&D carried out by semi-public R&D laboratories, and R&D carried out by universities and by other public R&D laboratories (e.g. hospitals and museums).

In the figure 6.2 national R&D has been divided into these five different categories for the years 1981 and 1989 for the three Nordic countries, and the totals have been given for a few other OECD countries.

If we start by looking at the levels and growth of R&D expenditure as a percentage of GDP, the Swedish level was considerably above the general OECD level in both years. In 1981 it was below Japan and the US whereas in 1989 it was at the same level as these two countries. Finland has had a very rapid growth in R&D in the eighties and it has almost caught up with Norway, the country with the second largest R&D performance in the Nordic region. By 1989 the Finnish level was

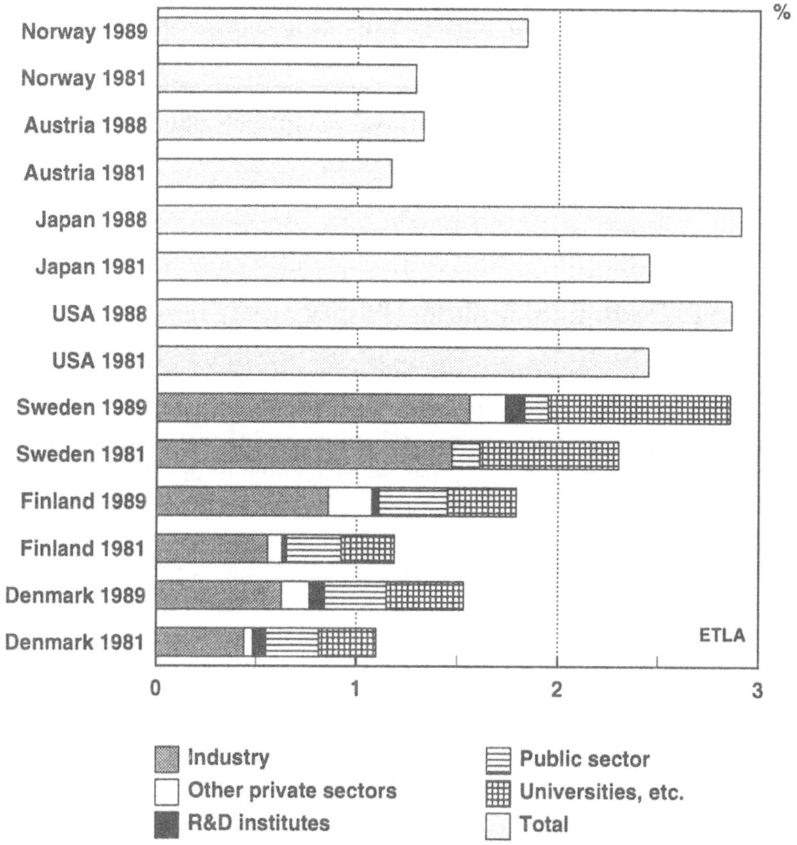

Fig. 6.2. The structure of R&D in Denmark, Finland and Sweden. The level of R&D (R&D/GDP, %) in selected OECD countries 1981 and 1988/89.
Note: The Swedish figures for 1981 do not allow a splitting of Industry, Other private sectors and R&D institutes. They are all included in Industry.
Source: Nordisk Industrifond 1991a and OECD 1991.

above the OECD average[9]. Although, in the eighties, growth of Danish R&D was higher than the OECD average, the level was still a bit below the OECD average in 1989. This is due to a very low level of R&D in the beginning of the period.

We can go into more detail of the structure of the R&D efforts in Denmark, Finland and Sweden if we look at the sectors which performed the R&D. In Sweden and Finland, approximately half of total R&D was performed by industry (in Sweden even 54% in 1989). This figure was about 40% for Denmark. The size of other R&D expenditures in Finland and Denmark was of the same magnitude, and this

[9] The OECD average was app. 1.6%

indicates that the reason for a low R&D/GDP ratio in Denmark (below the OECD average) is that industry has relatively little R&D.

If we carry the analysis of industrial R&D a bit further and focus on which branches the different countries specialize in, it turns out that Sweden is relatively specialized in 'high-tech' sectors[10] (2/3 of total R&D). Denmark is specialized in 'high-tech' sectors (53%) and 'medium tech' sectors (32%) while Finnish R&D is more spread over all sectors (Nordisk Industrifond 1991a). Furthermore, we can look at the volume of R&D performed by firms of different size in the three countries. This has been done in Vuori and Vuorinen (Introductory Chapter, table 1.5), and we see that a remarkably large share of R&D is carried out by small firms in Denmark compared to Finland and Sweden and this may explain why industrial R&D is so relatively low in Denmark. Norway, which has an industrial structure similar to the Danish, has the same features as Denmark with respect to its R&D structure.

When we focus on R&D performed by other private sectors and the public sector, we find that it is not just the level of R&D in other sectors than manufacturing that are similar in Finland and Denmark, but also the composition of this R&D. Universities and other public institutions each perform 20-25 per cent of total R&D. In Finland other private sectors account for a larger part of R&D than in Denmark, but R&D institutes account for a larger part of R&D in Denmark than in Finland. However, as we shall see later, this does not imply that the formal institutional set-up of the non-industry R&D is identical.

When we turn to Sweden we find a somewhat different composition of non-industry R&D. Universities, etc. account for approximately 70% of this 'other' R&D while other parts of the public sector only account for 11% of non-industry R&D. We shall return to this later.

The Education System. Another important element in the institutional set-up is the amount and organization of education and vocational training, and this and the following sub-section provides a short description of these systems in the three Nordic countries.

In the Introductory Chapter by Vuori and Vuorinen we find a comparison of the level of education in the Nordic countries. We may add three aspects to this analysis. First, it should be observed that the figures in table 1.3 are high compared to many countries outside the Nordic area (e.g. Great Britain), and this is an important feature of the Nordic welfare system. Second, that the *systems* of education are very similar in the three countries. They are all built up around a 9-10 year comprehensive school, a 2-3 year secondary school and a 3-5 year third level school.

Third, we shall argue that it is not just important for the innovation activities that the country has a large number of science and engineering graduates. Graduates

---

[10] According to OECD definitions, 'High tech' sectors are defined as sectors in which R&D expenditure exceeds 4% of turnover. 'Medium tech' sectors spend between 1 and 4% of turnover, and 'Low-tech' sectors spend less than 1% of turnover on R&D.

from the third level in general have a positive impact on innovation because firms need qualified personnel to cope with management, marketing, etc. in order to make successful innovations. In this respect Finland is doing quite well; in 1988 14.9% of those in the same age-group graduated from university compared to 10.9% in Sweden and 12.4% in Denmark (Year-book of Nordic Statistics 1992).

Continuing Vocational Training. During the eighties it was realized in most Western countries, including Denmark, Finland and Sweden, that vocational training did not match the demand for qualified labour. Demand had shifted, partly because firms had recognized that technological development is important as a factor of competition, and partly because governments had realized that a well-trained work-force was important as a generator of economic growth.

According to Dahlberg (1989), Sweden was the country which, in 1989, spent the largest proportion of GDP on training in relation to unemployment; almost 0.5% of GDP with an unemployment rate of 3%. Finland spent about 0.25% of GDP with an unemployment rate of 5%, and Denmark spent a little below 0.3% of GDP but with an unemployment rate of 6.2%. There seems to be a relationship between the extent of training and the unemployment rate: The bigger the unemployment rate, the smaller amount spent on training, and vice versa. The relationship appears even clearer if one compares the Nordic countries with other European countries.

Relationships Between Universities and Firms. Universities are often regarded as repositories of public knowledge; they draw on it in their teaching and they add to it through their research (Nelson 1986). Thus, the role of the universities in the economy is to secure basic research (open up new paradigms and trajectories) and to educate graduates. To make commercially directed applied research is not one of their main purposes. However, still more countries recognize that the large amount of money spent on universities (as shown in figure 6.2 between one fifth and one third of total R&D spending in the Nordic countries) could be better utilized if there were closer links between universities and industry.

Therefore both Denmark, Finland and Sweden have set up so-called science parks in relation to almost all their universities, and it is hoped that this sort of co-operation will result in a better transmission of scientific knowledge from the universities to the firms and thereby increase the technological opportunities for the firms.

The Nordic Innovation Survey (Nordisk Industrifond 1991b) is one of the few attempts made to produce comparable data on the extent of co-operation between universities and firms. In this survey contacts between universities and firms do not necessarily involve science parks. It is concluded that Swedish firms are involved in co-operation with universities more often than Danish and Finnish firms (46% compared to 35% and 34%, respectively). This may be one result of the Swedish concentration of public R&D on universities.

The Technological Service System. The next formal institution directly supporting technical change we want to deal with is the technological service system[11]. The aim behind the establishment of a national technological service system is to enhance the competitiveness of the industry as a whole and to secure an even development in different regions. Such technical service and information centres may be especially important for the ability to create and maintain innovative activities in small firms emphasizing incremental innovations.

At the 'government level', all three countries have administrative bodies that advise the government in industry- and technology-related areas[12] and support development of special industrial sectors or technology areas through grants and loans. At the 'service institute level' all the Nordic countries have a series of research institutes. In Denmark and Sweden these institutes are all small or medium-sized whereas the Finnish service sector is dominated by one large research institute (VTT).

Another major difference in the organizational set-up of the technological service system between Denmark on the one side and Finland and Sweden on the other is that the majority of Finnish and Swedish service institutes are headed and funded directly by government bodies, whereas the technological service system in Denmark is almost exclusively independent and self-financing.

The Financing of Innovation. One of the most important conditions for innovation is the possibility of financing it. Innovation is a process over time in which decisions have to be made in anticipation of future returns. Compared to normal investment decisions, however, the element of uncertainty is even more important. This makes the intermedaiary role of the financial system crucial in relation to innovation.

Clearly, there are important differences between national systems of innovation in this respect. It is not only a question of getting finance at a lower cost in one country compared to another. It is also a question about how dependent innovating firms are on external finance as compared to internal financing, which is related to structural factors, for example the firm size distribution in the country - large companies being less dependent than small firms.

It is also a question of institutional characteristics of the financial system, for example the division of labour between different institutions, the degree of concentration and centralization and its relations to the non-financial sector and to the government. Financial systems can be grouped into different categories. It is, for example, possible to make a distinction between capital-market based systems (the US and the UK are examples of this) and credit-based systems in which mainly banks finance innovation. Credit-based systems, in turn, can be divided into

---

[11] This is a very complex area, and impossible to describe in just a few lines, so this section only gives some very rough descriptions. Several other chapters in this volume include this issue, so we refer to these chapters for further discussion.

[12] In Denmark Industri- og Handelsstyrelsen, In Finland TEKES and in Sweden NUTEK.

systems with strong (Japan and France) and with weak (Germany) government regulation (Christensen 1992).

Different financial systems influence innovation processes in different ways. It can, for example, be argued that they differ in their ability to influence and support selective and lasting borrower-lender relationships in the system and that this is of special importance for the innovation process (Christensen 1992)

Denmark, Finland and Sweden can all be said to have credit-based systems of financing innovation, with strong borrower-lender relationships between banks and firms, and with moderate or little government intervention. But there are differences too. Financing through foreign banks and capital markets have been more important in Sweden than in the other countries. This is related to the relative importance of multinational firms.

Generally, large firms have the least difficulties in getting innovation activities financed, at least in the Nordic countries. Large firms, however, seem to have more financing problems in Finland than in Denmark and in, especially, Sweden. On the other hand, small firms seem to have relatively more problems in Sweden than in the other countries (Nordisk Industrifond 1991b). In Sweden large companies have also often played an important role in financing the carrying through of innovation projects started by small firms, for example by overtaking the firms. This has not, for obvious reasons, been common in Denmark and is probably not very important in Finland either. It also seems as if private investors play a more important role in financing innovation projects in Sweden than in the other countries.

The role of finance in the innovation process must, clearly, be characterized as an area very much in need for further research. We know that it is very important, which was already emphasized by Schumpeter, and that institutional change has been fast and comprehensive for some years now, but our knowledge is very limited both on the theoretical and empirical level.

6.4.2 Informal Institutions

We have defined institutions as the sets of habits, routines, norms and laws which regulate the relations between people and, thus, shape human interaction, and we have made a distinction between formal and informal institutions. It follows directly from this definition that interactive learning and innovation depend not only on formal institutions like R&D departments and so on but also, and fundamentally, on broad parts of the informal institutional set up. This becomes crucial, especially for incremental innovation as a normal and ubiquitous process in economies, where innovation has become an important form of competition.

Communication economies are central aspects of national systems of innovation and are to a large extent determined by the informal institutional set up. Communication and interaction inside firms depend on many aspects of work organization and management. Several authors (Aoki 1986, 1990, and Freeman 1987), have underlined the importance of institutionalized communication between procurement departments, production departments, marketing departments and R&D departments and other types of established horizontal information flows.

Interactive learning may be seriously hampered if, for example, the norms and habits of workers make them reluctant to communicate and cooperate with other 'levels' of the firm. Factors like trust and legitimacy, which may depend on institutional factors like participation and job security, as well as suitable procedures for reaching compromises, are important here. Traditional barriers between different skill groups and conflicts over the distribution of power and income, both in individual firms and in society at large, tend to make communication more difficult.

Communication and interaction between firms and other types of out-of-firm communication and cooperation feed much of the learning needed for innovation. Product innovation is more difficult without a feed-back of user-experiences. The involvement of suppliers in manufacturing processes gives them a better idea of the equipment they are to deliver, which facilitates process innovation. The Nordic Innovation Survey (Nordisk Industrifond 1991b) shows, that customers are an important source of product-innovation ideas in Scandinavian firms. Universities and R&D institutions are also frequently mentioned in this connection. The ability to establish, break up and re-establish out-of-firm relations and to cooperate in more or less informal networks for knowledge exchange seems to be essential for technological dynamism.

There are also many factors in the institutional system outside the enterprise sector, such as different social norms, which are important for interactive learning and innovation. One might mention norms about conflict, consensus and cooperation at all levels of the economy which influence many institutions of economic and industrial democracy, labour market institutions, corporatist institutions and so on. The degree of 'opportunistic behaviour' in the relations between different agents is also important.

It is easy to see that most of these institutional factors differ very much even between so closely related countries as the Nordic ones. It is much more difficult to identify the most important of these factors and measure their contributions to innovation performance. We lack both methods and data and an improvement of the situation requires a multi-disciplinary endeavour.

However, all three countries have both strong and weak elements in their institutional set-up. In Denmark several consumer-goods producing industries, for example furniture and clothing, have been able to repeatedly improve the products and establish a reputation for good design and good quality. This requires more or less continuous and low friction communication within and between often rather small firms. From visiting small and medium-sized firms you can easily get the impression that the 'cultural distances' between the different kinds of people participating in innovation processes are often quite short.

The survival of small-scale production building on artisan traditions has led to rather special work relations in the firms in which cooperation between owner and workers is often easy. It also seems as if craftsmanship traditions, and cooperation within as well as between firms, in some instances has led to innovations which normally would be forthcoming only in big enterprises. Examples of this can be found in the production of mobile telephones and wind-mills.

It seems as if effective communication economies at least partly, have compensated for some of the weaknesses in the Danish system of innovation, as for example the lack of large R&D-intensive big firms, and active industrial and technology policy.

In Sweden communication economies have different and in some respects more formalized foundations than in Denmark. They have been influenced by the relatively centralized and concentrated production structure in Sweden and by the fact that labour market organizations are relatively centralized and strong on both sides. Furthermore, state regulation of labour market relations and a welfare state with a strong, and until recently quite effective, commitment to full employment have stimulated employers to build up long-term relationships with workers and develop competence. The Swedish model has been a kind of 'consensus economy' with positive attitudes toward technical change. Shortages of labour have led to efforts of organizational design in order to increase labour productivity and decrease labour turn-over rates as, for example, in the Volvo Uddevalla factory. In conclusion, then, Sweden and Denmark represent rather different solutions to the problem of interactive learning and innovation.

It is important to note, however, that this experimental and at the same time consensus-seeking attitude toward technical and organizational change can disappear rather rapidly in periods of industrial crisis and restructuring. Employers have withdrawn from most of the formal cooperation with the unions and the state and the Uddevalla factory will probably be closed.

Finland can probably be said to lie between the Swedish and the Danish model, but closer to the Swedish one. Even if class conflicts earlier were comparatively severe in Finland the consensus-building tendencies have later been strong, especially after the Second World War.

The *incentive system* (work incentives, saving incentives, investment incentives, etc.) can be said to lie on the border between the formal and informal institutions. The tax system and income transfer system, for example, influence the pecuniary incentives and are parts of the formal institutional set-up. On the other hand, some labour market traditions, work norms and ideologies about fairness of remuneration, which belong to the informal institutional set-up, also have incentive effects.

In all Nordic countries many politicians and social scientists have argued that the effectiveness of the incentive system has been damaged by the development of the welfare state. If this is true, it may have had negative effects on the specific incentives for learning and innovation as well. However, the incentives to engage in interactive learning and innovation are of a rather complex nature and the importance of individualized pecuniary incentives may be overstated. In fact, they may weaken the engagement in open communication with other parties, foster opportunism and increase communication costs.

Innovation processes often lead to structural change with unevenly distributed social costs. This may provoke resistance to change and many institutions related to the welfare state, for example systems for income redistribution, social security policies, labour market policies, retraining policies and so on, which compensate the 'victims' of structural change and make it easier for them to move into new activities, may actually strengthen the incentives for innovation. Incentive systems that reinforce communication, co-operation and interactive learning are not simple to design, however, and the positive as well as negative innovation incentive effects of different aspects of the Nordic welfare states should be seriously investigated rather than taken for granted on ideological grounds.

6.5 Comparison of Innovation Policies

In section 6.1.2 we stated that innovation policies are interpreted very broadly in this paper. They include, of course, policy measures employed to influence innovation directly but they also include other types of policies that are not directly aimed at innovation activity. This second type of (innovation) policy may even be more important than the 'normal' type since it, in the long run, influences both the structural and the institutional set-up of the economy and the composition and level of demand, thereby changing the basis and incentive for innovative activity. We also argued that the distinction between innovation policies and the institutional set-up is blurred. This is reflected in the following section inasmuch as it covers some of the same sub-topics as section 6.4.1.

Since the emphasis in this chapter is on positive rather than on normative theory, we shall keep this section very brief and just make a presentation of possible innovation policies and discuss some general trends in innovation policy in Denmark, Finland and Sweden. The number of possible innovation policy interventions is very large, as depicted in table 6.7.[13]

Since innovation depends on a successful combination of technology supply and market demand, innovation policies can be divided into measures influencing these two 'sides' (this is, e.g. done by Rothwell and Zegveld 1981 and Porter 1990). The supply side includes for example grants, loans and subsidies, research laboratories, technical education and information networks and centres. The technical upgrading of production which has occurred since the introduction of computer aided production has demanded that all industrial countries have had to build up a 'technological infrastructure' and steer general support to technological development in order to keep pace in the 'international technology race'. Therefore most countries (including Denmark, Finland and Sweden) have employed such initiatives during the last one or two decades.

[13] Cf. for instance OECD (1992), *Science and Technology Policy - Review and Outlook*, for more detailed country specific describtions.

Table 6.7. Classification of government policy tools

Policy tool	Examples
1. Public enterprise	Innovation by publicly owned industries, setting up of new industries, pioneering use of new techniques by public corporations, participation in private enterprise
2. Scientific and technical	Research laboratories, support for research associations, learned societies, professional associations, research grants
3. Education	General education, universities, technical education, apprenticeship schemes, continuing and further education, retraining
4. Information	Information networks and centres, libraries, advisory and consultancy services, databases, liaison services
5. Financial	Grants, loans, subsidies, financial sharing arrangements, provision of equipment, buildings or services, loan guarantees, export credits, etc.
6. Taxation	Company, personal, indirect and payroll taxation, tax allowances
7. Legal and regulatory	Patents, environmental and health regulations, inspectorates, monopoly regulations
8. Political	Planning, regional policies, honours or awards for innovation, encouragement of mergers or joint consortia, public consultation
9. Procurement	Central or local government purchases and contracts, public corporations, R & D contracts, prototype purchases
10. Public services	Purchases, maintenance, supervision and innovation in health service, public building, construction, transport, telecommunications
11. Commercial	Trade agreements, tariffs, currency regulations
12. Overseas agent	Defence sales organizations

Source: Rothwell and Zegveld (1981), p. 61.

The demand side includes factors like public procurement and prototype purchasing. This type of policy has also been used in the Nordic area. An early example was the agreement between the Swedish government and L.M. Ericsson about the development of telecommunication services (cf. section 6.3.3). A recent example is the agreement between the Danish national railways and Scandia Randers about the development of a new type of train.

Although innovation policies in the Nordic region look similar at first sight, you will nevertheless probably be able to find important differences once you engage in a more thorough comparative analysis. This is not possible in this chapter, though.

Some of the policies in table 6.7 are more direct in their effect on innovation activity than others. For example, grants and loans, research laboratories and information networks and centres influence innovation directly (although it may be difficult to control the extent, speed and direction of this influence). Other policy measures, for instance technical education, universities, monopoly regulations and tax policies, influence the innovation process less directly.

Still other policies influence innovation more indirectly. These include general education, infrastructure and currency regulations. We believe, however, that there is yet another group of (general) policies, not included in the table, which influence innovation through their influence on communication and interactive learning and the level and composition of demand. They include employment policies and social security policies. Employment and social security influence the attachment to the labour market, workers' attitude towards introduction of new technology and organizational changes.

We do not regard direct and indirect innovation policies as separate groups of policies, but rather as extremes on a scale ranging from simple measure influencing selected parts of the innovation process directly to general policies changing the basis for interactive learning. It is obviously an enormous task to describe how innovation processes are affected by government policies. We do believe, though, that a reconsideration of the whole field of innovation policies in the light of this conception is needed.

6.6 Comparison of 'Performance'

In this section we focus on 'immediate performance' and 'ultimate performance' of the three innovation systems. These terms were, briefly, discussed in section 6.1.2 above.

6.6.1 Measurement of the Immediate Performance of the Nordic National Systems of Innovation

It is well known that there is a lack of good measures of immediate innovation performance and that further research is much needed in this field. In order to be able to make precise comparisons of performance between national systems of innovation we, ideally, need several kinds of indicators, for example indicators for different kinds of innovation (product innovation, process innovation, organizational innovation, etc.). We also need to distinguish between incremental and radical innovation, between activities in different phases of the innovation process ('invention-innovation-diffusion' aspects), between innovation in different industries, in different networks and user-producer constellations of firms, in different firm types, etc.

In some contrast to the need for a diversified set of different immediate performance indicators we choose the number of patent grants in the US as our main indicator in this section. In spite of the general lack of good indicators this is not the only possible choice. A number of alternative measures have been used. These include bibliometric studies, studies of technological balance of payments, so-called innovation studies and studies of trade in high-tech products. These indicators are not without problems, however. The share of high-tech products in foreign trade is not a direct measure of immediate innovation performance, for example (see section 6.3.3). It is rather part of a description of the economic structure, which may to some extent be a result of earlier, and influence later, innovation activities.

Rejecting these measures and choosing patent statistics do not reflect that this measure is without methodological weaknesses, but we shall not engage in a lengthy discussion about these[14]. Instead we shall use patents granted in the US as our patent measure to avoid some of the more serious methodological weaknesses.

Figure 6.3. below shows the number of patents granted in the US per million inhabitants for selected OECD countries. Concerning the first period, 1981-1984, the countries fall in three groups: high performance countries (Japan, The Federal Republic of Germany and Sweden, between 80 and 95 patents per year), medium performance countries (France, Great Britain and Austria, between 35 and 40 patents) and low performance countries (Finland, Denmark and Norway, between 18 and 28 patent grants per year).

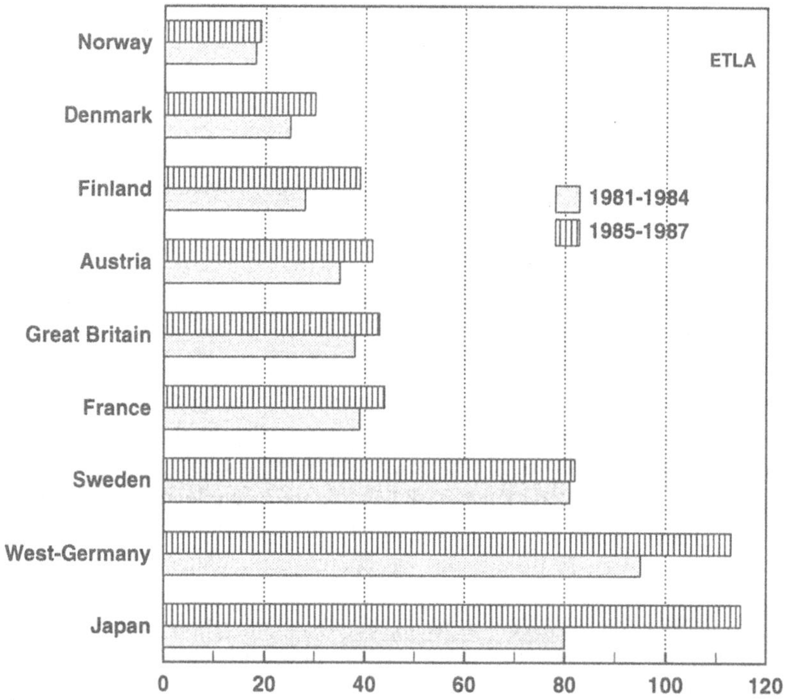

Fig 6.3. Average number of patents per year granted in the US per million inhabitants 1981-1984 and 1985-1987. Source: Science and Technology in Finland 1989.

[14] Basbjerg (1987) has an excellent discussion of these problems.

In the second period this picture has changed somewhat. If we focus on our three Nordic countries, Sweden has fallen behind Japan and the Federal Republic of Germany and Finland has caught up with France, Great Britain and Austria. This may indicate that, although it still produces comparably many patents, the Swedish system of innovation has run into difficulties in the late middle part of the eighties. Conversely the Finnish system of innovation, starting from a low level, seems to have grown stronger. The Danish system is performing rather poorly: patenting is at an internationally low level and growth is modest.

If we carry the analysis a bit further and focus on the number of patents in different product groups (figure 6.4) it turns out that Sweden performs best in all the aggregated IPC classes. There are only a few product groups in which Finland and Denmark are granted more than half the number of Swedish patents. In textile and paper, etc. Finland is granted almost as many patents as Sweden. In chemistry and metallurgy, etc. both Finland and Denmark are granted relatively many patents compared to Sweden and in food, etc. the Danish share is relatively high.

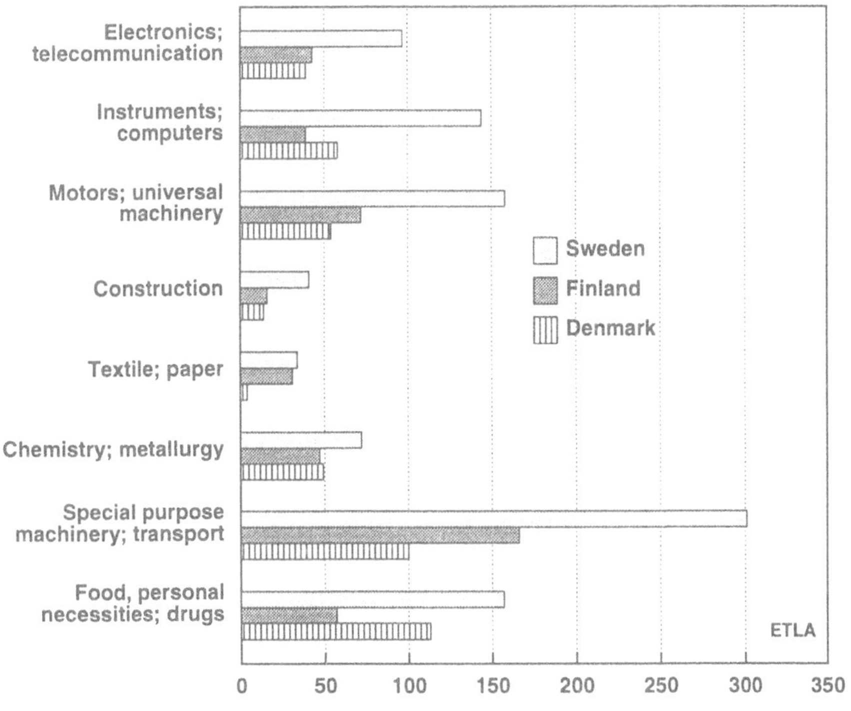

Fig 6.4. Number of patents granted in the US 1989 per million inhabitants. Distributed on IPC classes.* Source: Nordisk Industrifond 1992.

* See International Patent Classification for a detailed description of the IPC classes.

The method above is suitable for describing differences and similarities *between* countries but it cannot be used for a discussion of the relative strength of branches *within* the country since it reflects, at least to some degree, the different propensities to take out patents in different branches. Therefore, if we want to go into more detail about the strength of different branches within the individual country, we have to use a different method.

Figure 6.5 below represents one way to do this, namely the relative frequency of patents. The patenting of the manufacturing industry as a whole has the value one,

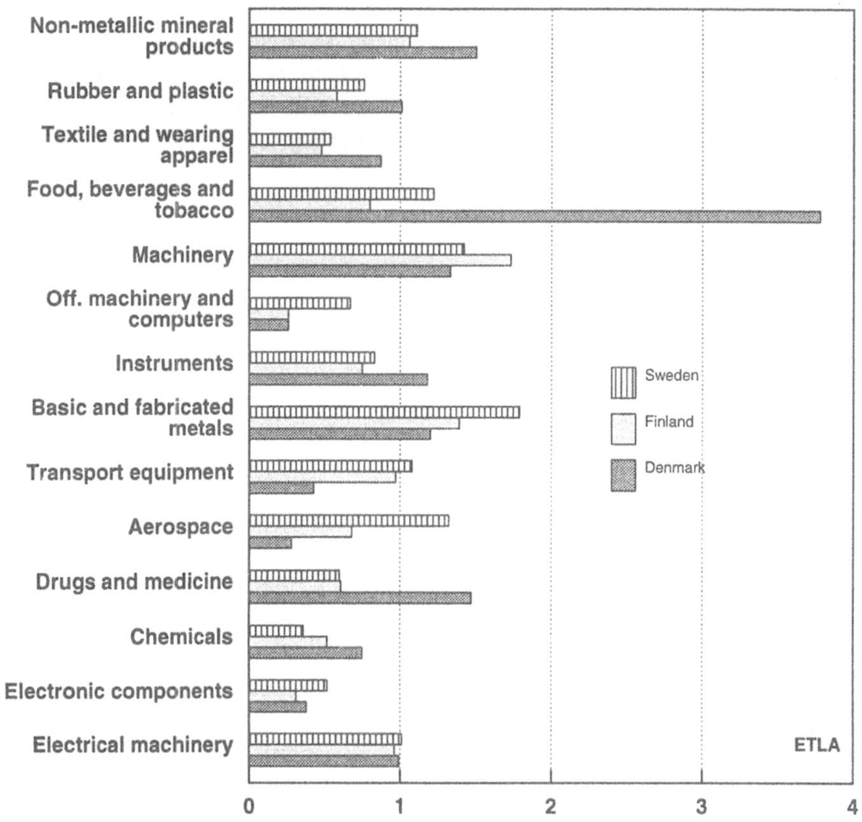

Fig. 6.5. The relative frequency of patents in the US in selected product groups 1975-1988.** Source: Nordisk Industrifond 1992.

** The relative frequency of patents by product group in the US for a given product group is defined as a country's share of patents in the US in this product group divided by the country's share of the total number of US patents for all manufacturing industries.

and values above this indicate a specialization in the area, and values below indicate the opposite[15].

The most conspicuous tendency in figure 6.5 is the Danish specialization in food, beverage and tobacco; within this branch a patent is granted almost four times as often (relatively) as in other branches in manufacturing industry. Also in non-metallic mineral products, machinery and drugs and medicine Danish firms take out a relatively large number of patents. Sweden primarily specializes in basic and fabricated metals and to some extent in aerospace and machinery. Finland is mainly specialized in machinery and in basic and fabricated metals. In both countries, however, the degree of specialization is much lower than the Danish specialization in food, etc.

These specializations in patenting reflect the different, historically determined industrial specializations in the three Nordic countries. As already mentioned in section 6.2, today's industrial structure is highly dependent on the individual country's (natural) resource base and the process of industrialization. Danish industry is particularly strong in activities connected to the agro-industrial development block (food, etc., dairy machinery, drugs and medicine). Swedish industry is strong in connection to metal ore and forestry (basic and fabricated metals and machinery) and Finland specializes in industries related to forestry (machinery and, although not included here, paper and pulp products).

Summing up, this brief analysis shows that Sweden is performing well, both at the total level (even though the growth seems to have stopped) and in all product groups; they are, however, especially strong in basic and fabricated metals, aerospace and machinery. Finland is catching up from the low-performance to the medium-performance group of countries, and specializes in machinery and basic and fabricated metals. Denmark is lagging behind these two countries, but the growth rate is higher than the Swedish one. They are highly in food, etc. and to some degree specialized in non-metallic mineral products, machinery and drugs and medicine. This reflects the existence of different 'development blocks' in the three Nordic countries.

6.6.2 Comparison of Ultimate Performance

There is, unfortunately, no simple and unambiguous measure or indicator which fully embrace the term 'social welfare' as an ultimate outcome of the innovation system. If, for instance, the 'ultimate performance' term is approached from a sociological angle, various measures as infant mortality rates, the amount of occupational diseases, frequency of labour conflicts, social security standards or environmental standards, would probably be included as relevant indicators. Political scientists might emphasize distribution of power among various citizens or the

[15] Cf. the note to figure 6.5.

degree of participation in central decision processes as important elements of the 'social welfare' variable.

It is possible to argue for many different types of indicators. It is important to emphasize, since it is the performance of innovation systems we are referring to, that they should relate to the dynamic efficiency properties of the economy. In this section, however, we use only traditional macroeconomic performance indicators as economic growth, unemployment, balance of payments and productivity, which already in the short run to some extent depend on innovation activities.

After the Second World War the basis for industrial development was quite different for the three countries. Sweden had the relatively most well-developed manufacturing sector of the three countries, while the manufacturing output per capita in Finland was well below the others. However, since the sixties and especially from the mid-seventies the manufacturing output has grown much faster in Finland than in the other two countries, cf. figure 6.6 below. Today, the relative size of the Finnish and the Swedish manufacturing sector is both around 25% of the GNP, while the Danish is only approximately 20%.

If we add developments of manufacturing employment to the picture, the differences become more evident. As figure 6.7 shows, both employment and output in Finnish manufacturing grew much faster than the OECD countries on average until the early eighties, while manufacturing employment in both Sweden and Denmark decreased more than the OECD countries. In the middle of the eighties the Danish economy experienced a brief period (1984-1987) of rapid growth in manufacturing output and employment combined with a decrease in the productivity level. In

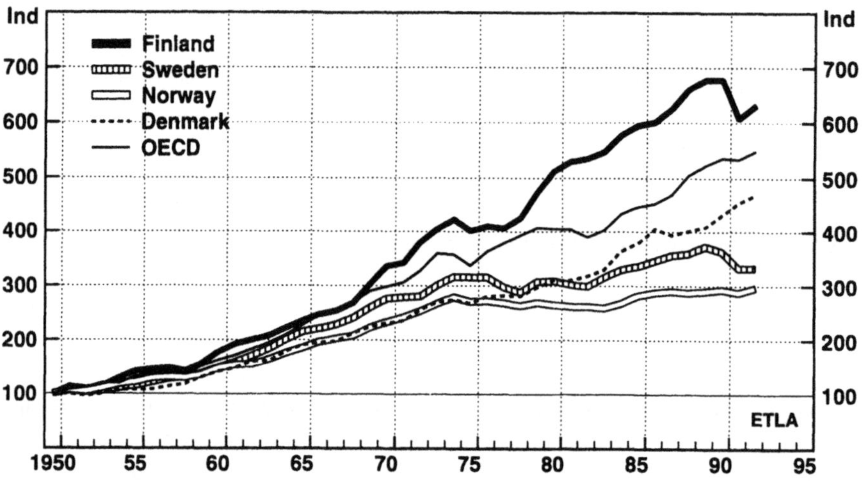

Fig. 6.6. Volume of manufacturing output in the Nordic countries and in the OECD (1950 = 100). Source: ETLA.

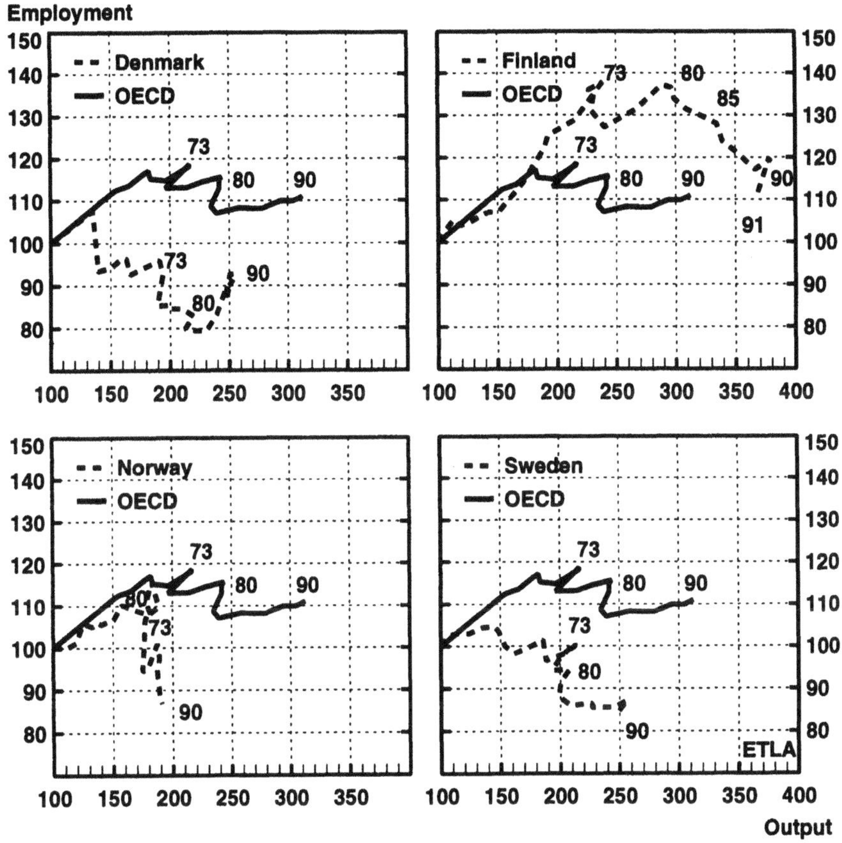

Fig. 6.7. Employment and output in manufacturing (1960=100). Source: ETLA.

Finland manufacturing employment fell, while it stabilized in Sweden in the same period.

The development in the unemployment rates, shown in figure 6.8, reflects another important difference in the 'ultimate economic performance' among the three countries. Using this indicator, Sweden has performed extremely well during the whole period, although, recently, the rate of unemployment has risen rapidly. The Danish rate of unemployment rose rapidly during the seventies and exceeded 10% in the beginning of the eighties. Unfortunately, the declining trend in the middle of the eighties was only temporary, and the rate of unemployment now (1993) exceeds 11%. Also in Finland unemployment increased during the seventies, but unlike the case of Denmark the development turned around in Finland and the unemployment rate in the 1980's stabilized at a fairly low level (about 5%). However, recently the rate of unemployment has exploded in Finland, from around 4% to 15% in a very brief period.

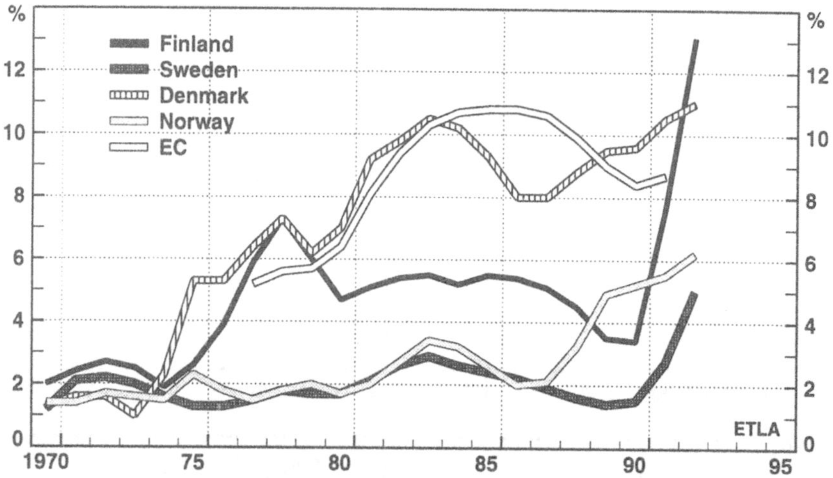

Fig. 6.8. Unemployment rate in the Nordic countries and EC, % of labour force

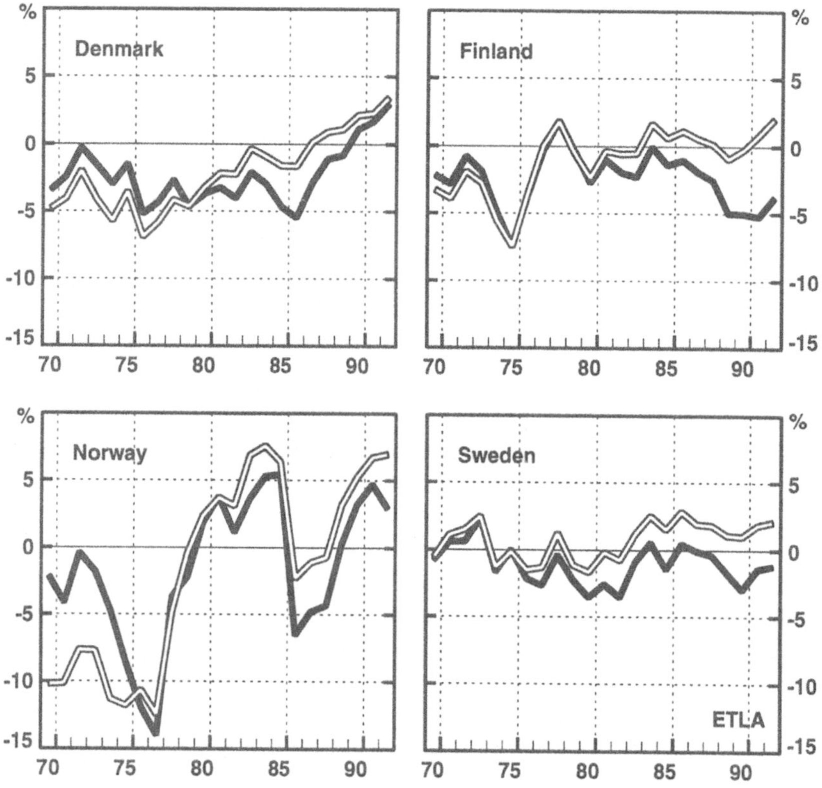

Fig. 6.9. Trade and current account, % of GDP. Source: National Accounts and Balance of Payment Statistics.

One of the main characteristics of a well performing national system of innovation is that it allows economic growth, and a high degree of freedom in economic policy, without an increasing international debt. The balance-of-payment restriction is especially severe for small and very open economies and all three countries have had these problems in varying degrees during the last few decades, as can be seen from Figure 6.9.

We are not able to discuss this or other 'ultimate' performance indicators more in the present context, however. The discussion in this section has mainly served to indicate the complexity of the problem of defining and measuring the performance of a national system of innovation. The value premises of innovation studies are often implicit or even hidden, but it should not be taken for granted that novelty and innovation are always good things and it seems important to make the value premises more explicit and to develop the performance indicators much more.

6.7 Conclusion

One of the purposes of this paper has been to emphasize the complex nature of a national system of innovation. A consequence of this is that comparisons between such systems must either involve quite a lot of empirical work or rely on some rather drastic selections from the set of relevant data.

Our approach to innovation theory has led us to concentrate on those aspects of the institutional set-up and the economic structure which most fundamentally shapes the processes of interactive learning. We have tried to argue that many institutional factors on the micro-, meso-, and macroeconomic levels affect the combination of different kinds of knowledge, communication economies and innovation. These factors range from the organization of the work process in the firm to the amount of communitarian elements in the culture of the country. Our selection in this paper, however, is not only a result of theoretical considerations. It is also influenced by statistical 'supply-side' factors and to some extent it is a shot in the dark. As economists, we do not feel especially well equipped for this task. It requires, we think, a multi-disciplinary endeavour.

When it comes to structural factors we think that the identification of, both existing and potential, industrial development blocks and the technological possibilities and bottle-necks, trajectories and communication patterns connected to them is crucial. These blocks should, however, not be analyzed in isolated national contexts only, but they should be compared between countries and related to trends in the international economic and technological development.

By concentrating on institutional and structural traits, the stability and durability of national systems of innovation are underlined. This raises the question of how such systems change, which we have not discussed in this paper. This question, obviously, becomes especially important in a period when a new techno-economic paradigm emerges in the global economy putting many national systems of innovation under strain.

An important aspect of this question is to what extent institutional 'design' and maybe even institutional borrowing from one country to another can become a part of economic policy. This possibility might even be considered as the main argument for comparative studies of national systems of innovation. But of course, we still need a much better description and mapping of the international institutional and structural diversity.

It must be admitted that our selection and comparison of empirical facts among the three countries do not permit any clear and solid conclusions about future trends in the Nordic national systems of innovation or about policies for affecting this development. We may be in a position, however, to raise a few relevant questions in relation to the need for further research.

How can it be, for example, that Denmark with so little R&D activities, so poor patenting records, so few large firms and with so many resources tied up in an agro-industrial block still can show reasonable economic growth records and remain in 'the rich countries' club? And for how long can this go on? We have indicated that the answer has something to do with elements in the informal institutional set-up affecting communication economies and interactive learning. Communication costs are low and craftsmanship traditions and norms about cooperation have in many parts of the economy transformed incremental innovation into almost a habit. This, however, remains largely an unresearched area.

For Sweden (and to some extent Finland) the question can be turned around. How can a severe industrial crisis develop so fast in a country with so much research and development, so many patents and so many large and advanced firms? Clearly, the connections between immediate and ultimate performance are not very stable. And it seems as if the ability of the institutional set-up to support dynamic efficiency can deteriorate rather quickly.

And more generally, for how long can all three countries defend their positions as rich countries while continuing to depend on raw material-related development blocks? Which changes in economic structure are most urgently needed? Which institutional changes would facilitate such structural change?

It seems as if the recent interest in national systems of innovation has opened up a vast new research area but that there still is a considerable way to go before we gain economically and politically useful results. Comparative studies of national systems of innovation are necessary elements in such a research endeavour.

References

Andersen, E. S. (1992), 'Approaching National Systems of Innovation', in Lundvall, Bengt-Åke (ed.) (1992), *National Systems of Innovation. Towards a Theory of Innovation and Interactive Learning*, London, Pinter Publishers.

Andersen, E. S. and Lundvall, B.-Å. (1988), 'Small Countries Facing Technological Revolutions: An Analytical Framework', in Freeman, C. and Lundvall, B.Å. (eds.), *Small Countries Facing the Technological Revolution*, London, Pinter Publishers.

Andreasen, P.B. & P. Lotz (1990), *The Research Laboratory "Denmark" - Innovation in Danish Health Care Industry*, Report prepared for the Danish Ministry of Foreign Affairs, Copenhagen.

Aoki, M. (1986), 'Horizontal vs. Vertical Information Structure of the Firm', *American Economic Review*, Vol. 76, No. 5.

Aoki, M. (1990), 'The Participatory Generation of Information Rents and the Theory of the Firm', in Aoki, M. et al. (eds.), *The Firm as a Nexus of Treaties*, London, New Park, New Dehli, Sage Publications.

Basbjerg, B.L. (1983), Patents and the Measurement of Technological Change, *Research Policy* vol. 16 no. 2-4.

Brændgaard, A., Gregersen, B., Lundvall, B.-Å., Olesen, N. M. and Aaen, I (1984), 'Strukturproblemer, Teknologipolitik, Sociale innovationer, *Serie om industriel udvikling* nr. 30, Aalborg Universitetsforlag.

Christensen, J.L. (1992), *The Role of Finance in Industrial Innovation - Aspects of Financial Systems*, Aalborg, Aalborg Universitets Forlag.

Central Statistical Office in Finland (1990), *Science and Technology in Finland 1989*, Helsinki.

Dahlberg, Å (1989), 'Utbildning för 1990-talets arbetsmarknad' in Mohlin, B (ed.) *Nordisk Konferens om Utbildning för 1990-talet*, Göteborg, Graphic Systems.

Dalum, B. (1992), 'Export Specialization, Structural Competi- tiveness and National Systems of Innovation' in Lundvall, B.-Å. (ed.), *National Systems of Innovation Towards a Theory of Innovation and Interactive Learning*, London, Pinter Publishers.

Dalum, B., Møller K. , Jørgensen U. and Valentin F. (1991), 'Inter- nationalisering og erhvervsudvikling', *Notat*, København, Industri- og Handelsstyrelsen.

Edquist, C. and Lundvall B.Å. (1993), 'Comparing the Danish and the Swedish Systems of Innovation' in Nelson, R.R. (ed.), *National Systems of Innovation*, Forthcoming.

Edquist, C. and McKelvey, M. (1991), *The Diffusion of New Product Technologies and Productivity Growth in Swedish Industry*, CCC Working Paper no. 91-15, University of California at Bakerly.

ETLA, IFF, IUI, NøI (1990), *Growth and Integration in a Nordic Perspective*, Finland.

Freeman, C. (1987), *Technology and Economic Performance: Lessons from Japan*, London, Pinter Publishers.

Freeman, C. and Lundvall, B.Å. (eds.), (1988), *Small Countries Facing the Technological Revolution*, London, Pinter Publishers.

Grandstrand, O. and Sigurdson J. (eds.) (1985), *Technological Innovation and Industrial Development in Telecommunications. The Role of Public Buying in the Telecommunication Sector in the Nordic Countries*, Lund, Nordforsk and Research Policy Institute.

Gregersen, B. (1992), 'The Public Sector as a Pacer in National Systems of Innovation' in Lundvall, B.-Å. (ed.), *National Systems of Innovation Towards a Theory of Innovation and Interactive Learning*, London, Pinter Publishers.

Heum, P. and Ylä-Anttila (1992), 'Firm Dynamics in a Nordic Perspective - Large Corporations and Industrial Transformation', *Discussion Paper*, ETLA, Finland.

Johnson, B. (1992), 'Institutional Learning', in Lundvall, Bengt-Åke (ed.) (1992), *National Systems of Innovation. Towards a Theory of Innovation and Interactive Learning*, Pinter Publishers.

Jørgensen, U. (1986), 'Elektronikbranchens etablering og strukturelle udvikling', *Forskningsrapport nr. 11, Institut for Samfundsfag*, DtH, Copenhagen.

Lemola, T. and Lovio R. (1988), 'Possibilities for a Small Country in High-Technology Production: The Electronics Industry in Finland' in C. Freeman and B.-Å. Lundvall (eds.), *Small Countries Facing the Technological Revolution*, London, Pinter Publishers.

Lovio, R. (1992), 'The influence of foreign companies on the birth and development of the Finnish electronic industry' in Vuori, S. and Ylä-Anttila, P., (eds.) (1992), *Mastering Technology Diffusion - The Finnish Experience*, Helsinki, ETLA.

Lundvall, B.Å. (ed.) (1992), *National Systems of Innovation. Towards a Theory of Innovation and Interactive Learning*, London, Pinter Publishers.

Lundvall, B.Å., Olesen N.M. and Aaen I. (1984), *Det Landbrugs- industrielle Kompleks*, Aalborg, Aalborg Universitstsforlag.

Maddison, A. (1991), *Dynamic Forces in Capitalist Development*, Oxford University Press.

Mjøset, L. (ed.) (1986), *Norden - Dagen derpå*, København, Samfundslitteratur.

Mjøset, L. (1992), *The Irish Economy in a Comparative Institutional Perspective*, Dublin, National Economic and Social Council, No. 93.

Nelson, R.R. (1986) 'Institutions Supporting Technical Advance in Industry', *American Economic Review*, vol. 76 no. 2.

Nordisk Industrifond (1991a), Nordisk FoU-statistik for 1989 og statsbudgettanalyse 1991, *Information*, no. 7, Oslo.

Nordisk Industrifond (1991b), Innovation activities in the Nordic Countries, *Information*, no. 3, Oslo.

Nordisk Industrifond (1992), Vitenskaps- og teknologi-indiaktorer for Norden, *Information*, no. 3, Oslo.

OECD (1991), *Basic Science and Technology Statistics*, Paris, OECD.

OECD (1992), *Science and Technology Policy, Review and Outlook 1991*, Paris, OECD.

Porter, M. E. (1990), *The Competitive Advantage of Nations*, New York, The Free Press.

Raumolin, J. (1992), 'The diffusion of technology in the forest and mining sector in Finland' in Vuori, S. and Ylä-Anttila, P. (eds.) (1992), *Mastering Technology Diffusion - The Finnish Experience*, Helsinki, ETLA.

Rothwell, R. & W. Zegveld (1981), *Industrial Innovation and Public Policy*, Frances Pinter, London.

Undervisningsministeriet (1991), *Udviklingen i forskning og teknologi i 1980'erne - belyst med videnskabs- og teknologi- indikatorer*, København.

Vuori, S. and Ylä-Anttila, P., (eds.) (1992), *Mastering Technology Diffusion - The Finnish Experience*, Helsinki, ETLA.

Year-book of Nordic Statistics 1992.

Appendix 1. Development Blocks

1. Agro-food	**(SITC)**
Meat & meat preparations	00, 01, 091.3, 411.3
Dairy products	02
Fish & fish preparations	03, 411.1
Cereals & cereal preparations	04
Feeding-stuff for animals	08
Other food products	05, 06, 07, 091.4, 099
Beverages & tobacco	11, 12
Animal & vegetable oil & fats	2, 42, 43
Cut flowers, bulbs & oth. plants	292.1-4, 292.6-9
Seeds and spores for planting	292.5
Agricul. & food proces. mach.	712, 718.3
Heating & cooling equipment	719.1
2. Forestry-paper	
Wood & wood maufactures	24, 63
Pulp & paper	25, 64
Paper & pulp machinery	718.1
Furniture	82
3. Textile-clothing	
Textile fibres	26
Textile yearn, fabrics, etc.	65
Textile & sewing machines	717.1, 718.3
Clothing	84
4. Metal	
Iron ore	281
Iron, steel & ferro-alloys	67
Aluminium	684
Non-ferrous ores & metals	282-86, 681-83, 685-89
Metalworking machinery	715
Manufactures of metal	69, 719.4, 812.1, 812.3
5. Health	
Pharmaceuticals	54
Electromedical equipment	726
Non-elec. medical equipment	861.7
Orthopaed. eq. & hearing aids	899.6
6. Telecommunication	
Telecommunications equipment	724.9

7. Rest: resource based	
Skins & leather manufactures	21, 61, 291
Crude fertilizers, crude minerals & coal	27, 32
Non-metallic minerals (cement, bricks, ceramics, glass, etc.)	66
Rest: rubber, electr. energy	23, 62, 35
8. Oil & gas	
Oil & gas	33, 34 s
9. Rest: chemicals	
Organic chemicals	512
Inorganic chemicals	513, 514
Dyestuffs, coloring materials	53
Fertilizers, manufactured	56
Plastic materials	581.1, 581.2
Other chemicals	515, 52, 55, 57, 581.3, 581.9, 59
10. Rest: engineering	
Machinery for other special industries or processes	717.2,718.2,718.4-5,719.3, 719.5, 719.8
Power generating machinery	711
Pumps & centrifuges	719.2
Typewriters & office machinery	714.1, 714.9
Computers & peripherals	714.2-3
Semiconductors	729.3
Machinery for production & distribution of electricity	722-3
Consumer electronics	724.1-2, 891.1
Domestic electronical equipment	725
Measuring & control. instruments	729.5
Photographic & optical goods, watches	861.1-6, 861.8-9, 862, 864
Railway vehicles	731
Road motor vehicles	732
Aircraft	734
Ships and boats (& oilrigs)	735
Other non-electrical equipment	719.6-7, 719.9, 733
Other electrical equipment	729.1-2, 729.4, 729.6-7, 729.9
11. N.e.s.	
Industrial products, n.e.s.	812.2,812.4, 83, 85, 863, 891.2-9, 892-97, 899.1-5, 899.9, 9

7 Technological Systems and Economic Growth: Comparing Finland, Sweden, Japan, and the United States

Bo Carlsson
Case Western Reserve University, USA

7.1 Introduction

The purpose of this paper is to compare Finland's technical and economic performance over the last 20 years to that of Sweden, Japan, and the United States within the context of a discussion of the Finnish national system of innovation, to comment upon the usefulness of the concept of technological systems as distinct from national systems of innovation, and to suggest areas for further research.

The paper is organized as follows. We begin by comparing the performance of the four countries in international trade, industrial structure and dynamism, industrial research and development (R&D), and patenting. Then follows a discussion of technological systems as distinct from national systems of innovation, their role in economic growth, their broad features, and the observable differences among the four countries in the technological system in one particular area, namely that supporting factory automation. In the concluding section, the findings of the study are summarized and some suggestions are offered concerning areas for future research.

7.2 International Trade Performance

Because of the difficulties inherent in using gross national product (GNP) as a measure of output, particularly in international comparisons, I will use international trade shares as a performance indicator instead.[1]

[1] The difficulties of using GNP as a measure of output arise primarily from the following considerations: GNP was originally designed to indicate the level of overall economic activity in the country on a short-term basis; it certainly does not indicate output, particularly in view of the fact that inputs rather than outputs are used to represent activity in the service sectors which make up by far the largest share of the economy; since the accounting and classification rules vary from one country to the next, international comparisons of GNP are particularly tricky; and, due to the index number problem, GNP does not adequately reflect the most important changes in the economy, namely those having to do with new and improved goods and services.

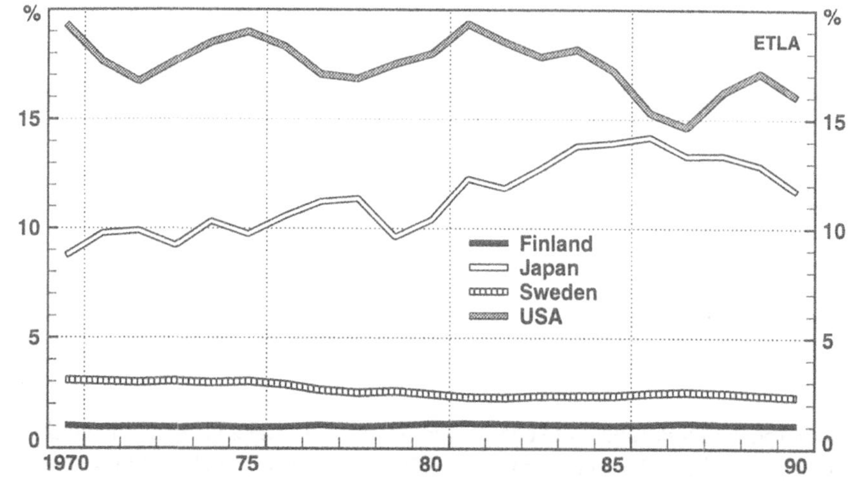

Fig. 7.1. Shares of industrial country exports. Source: IMF, International Financial Statistics Yearbook, 1991, pp. 120-1 and 124-5.

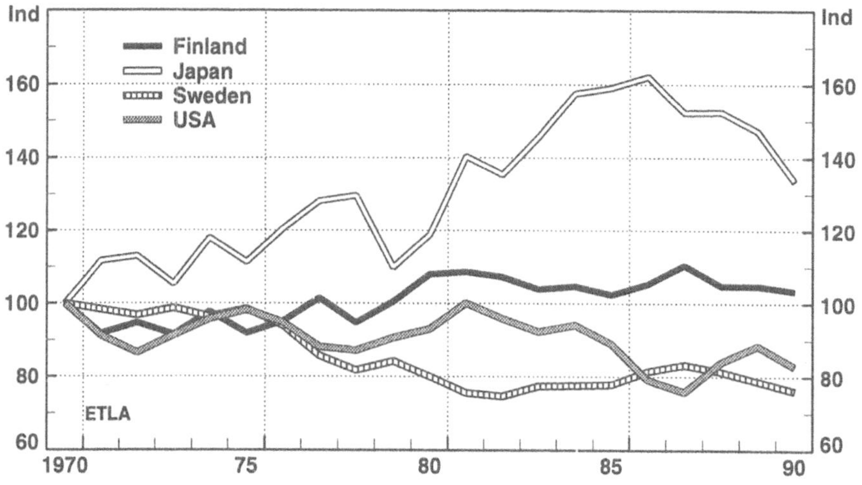

Fig. 7.2. Shares of industrial country exports, 1970=100. Source: IMF, International Financial Statistics Yearbook, 1991, pp. 120-1 and 124-5.

The measure of international trade performance most commonly used is the country's exports in relation to exports of other countries. In Figure 7.1, the exports of Finland, Sweden, Japan, and the United States over the period 1970-90 are expressed as a percentage of the exports of all industrial countries (as defined by the International Monetary Fund). The figure indicates that Finland has maintained its share of industrial country exports (about 1 percent) over the last two decades, while Sweden and the U.S. have been losing shares and Japan has been gaining. Figure 7.2 shows the same development in index form with 1970 as base year. According to this more detailed comparison, Finland lost nearly 10 percent of its "market share" in the early 1970s, regained and surpassed it in the early 1980s, ending up with a slightly larger market share in 1990 than in 1970. Sweden maintained its market share in the early 1970s (through the first oil crisis in 1973-74) but then lost about 25 percent in the late 1970s when the mining and steel industries, the shipyards, and the shipping business suffered substantial declines and in some cases complete elimination, and parts of the forest-based industries also experienced both cyclical and restructuring difficulties. (Carlsson et al. 1979, 18-23.) The United States experienced cyclical ups and downs during the 1970s and then a more prolonged decline during the 1980s, partly as a result of the rising value of the dollar. Meanwhile, the Japanese market share surged ahead, its growth sputtering during the oil crises of 1973-74 and 1979-80, then reaching a peak in 1987 but after that a rapid decline. The recent decline is presumably related to overseas investment by Japanese firms, resulting in offshore production replacing Japanese exports.

The increasing role of multinational firms is evident not only in the Japanese case but also, and even more so, in Sweden and the United States, as well as in Finland. As Blomström and Lipsey (1989) have shown, while both the United States and Sweden lost about 20 percent of their shares of world exports of manufactures between the mid-1960s and mid-1980s, the export shares of their multinational firms stayed fairly stable or even increased. The shares of U.S. and Swedish parent firms' exports as a percentage of total home-country exports also increased. In the Swedish case this occurred while the Swedish firms with production facilities abroad expanded their foreign production even faster than they increased their exports from Sweden (Swedenborg et al. 1988, 35). Furthermore, Encarnation (1992) has shown that the foreign sales by majority-owned subsidiaries of U.S. multinationals were approximately three times larger than U.S. exports in 1988 (six times larger in the European Community than U.S. exports to the EC) and in the case of Japan almost twice as large. Calculations for Sweden, based on material from Swedenborg et al., suggests that the corresponding Swedish figure is 0.7. The lower Swedish figure reflects the fact that the ratio of exports to total domestic manufacturing output is considerably higher than those in Japan and the United States, i.e., that *all* firms (not just multinationals) export a substantial share of their output. While comparable Finnish data are not currently available, the spectacular increase of foreign direct investment by Finnish firms in the latter half of the 1980s indicates a sharp increase in the degree of international integration of leading Finnish firms. (Kajaste, Parviainen and Ylä-Anttila, 1992.)

The implication here is that the world export share of a country may or may not reflect the country's international competitiveness, depending on how one views the role of its multinational firms. Surely a country is better off if it has domestically based multinational firms with strong positions abroad than it would be in the absence of such firms - provided that there are domestic spin-offs in the form of technology transfer, both direct and indirect domestic employment, and build-up of domestic infrastructure. What is important in the present context is simply that the technology base supporting the country's economic growth must be larger, the greater the role of domestically based multinational firms which draw upon that base not only for their domestic activities but also for their overseas production.

There are other reasons why export shares may not be a satisfactory indicator of international performance. Of particular interest here is that for varous reasons (e.g. exchange rate fluctuations reflecting real interest rate differentials and relative attractiveness to foreign investment because of market opportunities or availability of resources) a country may incur a trade surplus or deficit for quite prolonged periods. For example, both Sweden and the United States sustained sizable trade deficits for several decades during the 19th century; the U.S. had a surplus during most of the first six decades of this century; and Japan had a trade surplus during 17 years in the period 1970-90. (Carlsson 1990; IMF 1991.) In the U.S. case it might be argued that one of the reasons for the trade deficit during the 1980s was the relative attractiveness of the U.S. market for foreign investment in terms of both market opportunities (the U.S. was growing faster than Europe and had higher real rates of return) and avoiding the potential threat of protectionistic measures.

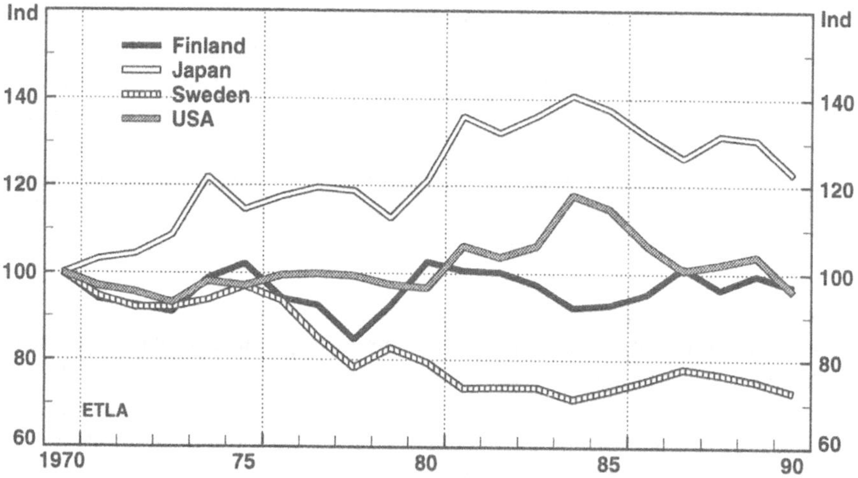

Fig. 7.3. Shares of industrial country trade, 1970=100. Source: IMF, International Financial Statistics Yearbook, 1991, pp. 120-1 and 124-5.

Thus, it may be of interest to compare the development of trade shares (the sum of exports and imports) of various countries, not just export shares. Such a comparison is made in Figure 7.3. The results are broadly similar to those in Figure 7.2: the Finnish share has fluctuated around a constant level, while that of Sweden has declined and that of Japan increased (although the Japanese increase is not quite as dramatic as in the case of exports alone). However, for the United States the large trade deficits during the 1980s resulted in a temporary *increase* in its trade share in the mid-1980s and then a stabilization around its 1970 level.

What does this mean in the present context? Broadly speaking, the United States attracted a lot of foreign investment during the 1980s, particularly from Japan but also from Europe. Partly as a result of this, its trade deficit started shrinking in the late 1980s as foreign firms relied more on production in their U.S. subsidiaries and less on exporting to the U.S. The Japanese export surplus began to shrink partly as a result of its overseas investments. Neither Finland nor Sweden has attracted much foreign investment. This is a problem to which we shall return. Swedish firms have invested heavily in Europe during the last several decades but have financed most of their expansion from foreign sources rather than via exports from Sweden. It appears that many of the largest Finnish firms have pursued the same strategy.

7.3 Industrial Structure and Performance

The economic performance of a country is necessarily tied to its history. One legacy of the past is the structure and composition of its industrial output. Even today the availability of forest-based natural resources plays a major role in both the Finnish and the Swedish economies.

Table 7.1 shows a comparison of the distribution of manufacturing value added by selected industry groups in Finland, Sweden, Japan, and the United States in 1970, 1980 and 1989. The table shows that the forest-based industries continue to play a major role in Finnish industry and that the engineering industries, although they are gaining rapidly, are still small in international comparison. This needs to be kept in mind when examining the Finnish economic performance. In Sweden, the forest-based industries have maintained their share, while the steel and textiles & apparel industries have declined and the engineering industries have gained. The composition of output is remarkably similar in Japan and the United States, except for the larger role of the steel industry in Japan.

At this highly aggregated level of analysis, it is difficult to discern major changes over time. At a more detailed level (comprising the same data broken down into 21 industry categories) it can be calculated that the weighted average percentage share change between 1970 and 1980 in Finland was 14.3 percent. This

Table 7.1. Distribution of manufacturing value added by industry in Finland, Sweden, Japan, and the United States, 1970, 1980 and 1989

		Finland			Sweden			Japan			United States		
ISIC	Industry	1970	1980	1989	1970	1980	1989	1970	1980	1989	1970	1980	1989
31	Food, beverages and tobacco	13.4	11.7	12.0	10.1	10.2	9.6	7.9	9.2	9.1	11.6	10.6	11.6
32	Textiles, clothing, leather & footwear	9.8	8.1	4.0	5.7	3.0	2.1	8.1	6.6	4.9	8.0	6.2	5.0
33+ 341	Forest-based products	26.2	24.7	23.3	16.7	16.7	16.1	7.2	6.5	5.2	6.9	6.8	7.4
35	Chemical products	10.0	10.8	11.1	8.0	9.8	10.4	12.9	12.5	12.6	13.7	14.2	14.5
37	Basic metals	4.6	4.8	5.7	9.1	6.6	6.1	9.4	10.0	7.1	6.7	5.9	3.9
38	Engineering industries	24.5	26.9	30.0	39.0	42.5	43.7	41.8	41.0	47.0	40.7	44.2	43.3
	Other industries	11.0	12.4	13.0	10.6	9.9	9.6	10.6	11.7	11.2	10.7	10.4	11.7

Source: United Nations, Industrial Statistics Yearbook, Vol. I, various issues.

compares with 15.7 percent in Sweden, 14.4 percent in Japan, and 13.1 percent in the United States. It thus appears that there were no major differences among these countries in the rate of change of structural composition of output during the 1970s, although there was a slightly more rapid change in Sweden and a slightly less rapid change in the United States. However, during the period 1980-89 the pace of change increased dramatically in Japan, Sweden, and the U.S. (to 18.1, 19.0, and 20.5 %, respectively), while it actually declined somewhat (to 13.8 %) in Finland. It is somewhat surprising that the composition of output did not change faster in Finland, given the country's late industrialization, the "catch-up" effect, and the common perception of rapid change in Finnish industry.

Other indicators of "industrial dynamism" are the changes in manufacturing output and employment. These are shown in Figures 7.4 and 7.5. Here the Finnish performance comes through more clearly. As far as manufacturing output is concerned, Finland (along with Japan) was the leading performer among the countries studied here over most of the period 1970-1989; only in the last two years was it surpassed by Japan. Manufacturing output more than doubled in Japan and Finland and almost doubled in the United States, while it grew only 30 % in Sweden.

Looking at manufacturing employment instead, Finland and the U.S. saw manufacturing employment increase during the 1970s and then decline in the 1980s, with the end result that there was virtually no change between the beginning and end of the period. In Japan and Sweden, manufacturing employment peaked in 1973 and 1975, respectively, declined in the late 1970s, and then held steady. In Sweden, manufacturing employment dropped by more than 15 percent between the early 1970s and the early 1980s.

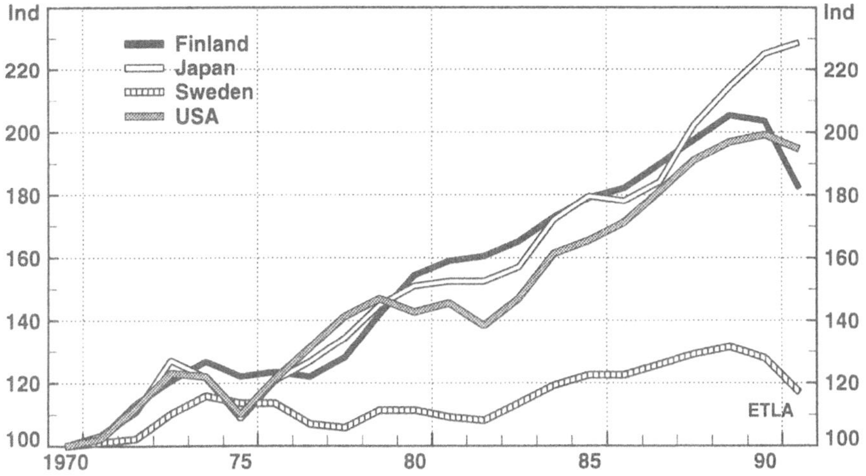

Fig. 7.4. Manufacturing output, 1970-1989. Source: United Nations, Industrial Statistics Yearbook, vol. 1, General Industrial Statistics, various issues; OECD.

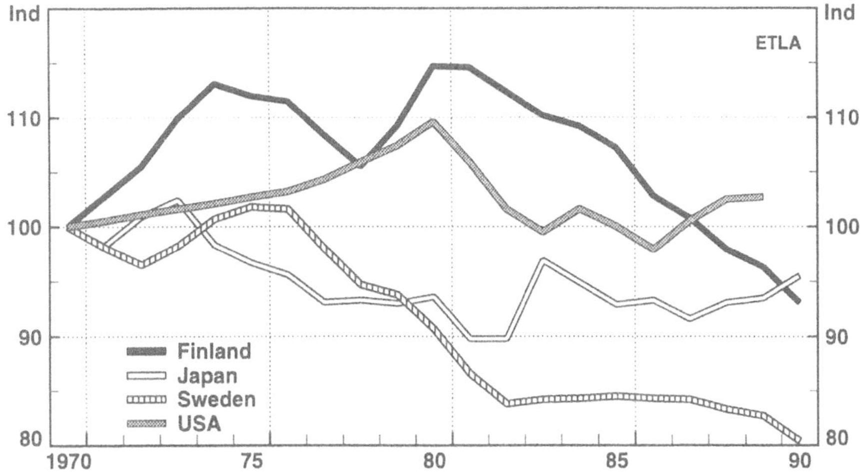

Fig. 7.5. Manufacturing employment, 1970=1989. Source: United Nations, Industrial Statistics Yearbook, vol. 1, General Industrial Statistics, various issues.

Another indicator of industrial dynamism is the rate of new business formation. While no internationally comparable data on this variable are currently available, a recent study (Carlsson 1992a) provides information on changes in the number of plants in manufacturing. A rapid growth in the number of plants presumably indicates a more dynamic industrial environment than a slower (or even negative) growth. (Ideally one would like to have information on *gross* flows rather than net, i.e. both entries and exits, not only their sum.) Figure 7.6 shows that the number of manufacturing establishments declined in Finland by about 10 percent during the first half of the 1970s and then grew by about 20 percent until 1985 and then fell again. As a result, the number of establishments grew only modestly in Finland over the period as a whole, while it increased by about 15 % in Japan and the United States. Again, Sweden is "the odd man out," showing a steady decline throughout the period. Indeed, a more detailed study (Carlsson 1992d) shows that the number of industrial establishments in Sweden fell from nearly 14,000 in 1968 to just over 9,000 in 1988. The most severe decline occurred in the smallest size category (less than 10 employees), from 5,831 to 1,234 establishments. Even more remarkably, the number of plants in this size category fell in *every* 2-digit NACE industry. This confirms evidence from other sources suggesting that there is a lack of entrepreneurship in Swedish industry and that this might explain at least part of the dismal Swedish industrial performance in the last 15 years.

There is another aspect of industrial structure which warrants comment. In their study of the national systems of innovation in Denmark and Sweden, Edquist and Lundvall (1989) point out the importance of agricultural cooperatives in explaining the present size and ownership structure of Danish manufacturing firms. The cooperative movement effectively prevented the accumulation of private capital and therefore led to a relatively unconcentrated industrial structure in Denmark. In Sweden, on the other hand, the early industrial development was based on forestry and mining rather than agriculture, requiring more capital and technology as well as larger enterprises. In addition to imported technology and expertise, the

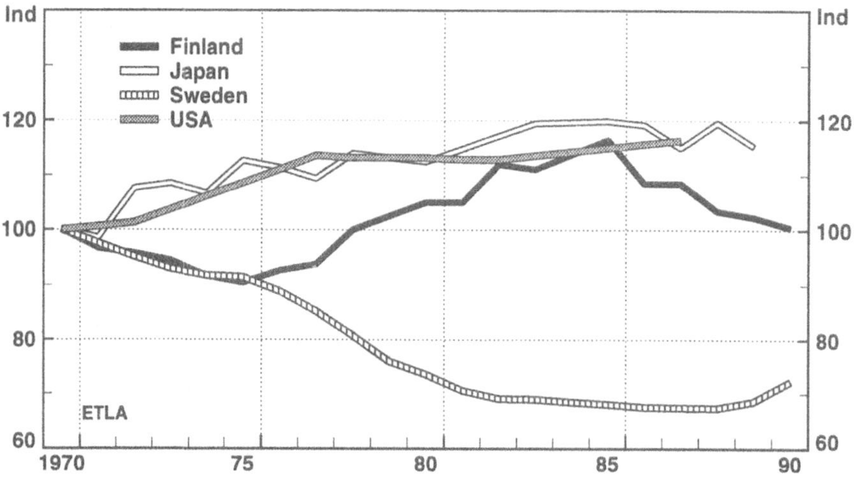

Fig. 7.6. Number of manufacturing establishments, 1970=100. Source: United Nations, Industrial Statistics Yearbook, vol. 1, General Industrial Statistics, various issues.

emerging Swedish engineering industry was based upon a number of specific innovations by Swedish entrepreneurs (ball bearings, telephones, light houses, separators, high voltage electric power transmission, etc.). Therefore, the Swedish industry became highly concentrated in the hands of a few families (capitalists and entrepreneurs) with close ties to certain banks. Thus emerged a structure of ownership and managerial control which bears strong resemblance to the pattern observed in Japan, the *zaibatsu* system which was replaced by *keiretsu* after the Second World War.[2]

The Swedish system seems to have been as successful in generating growth and allocating resources at least through the early 1970s as the Japanese system has been up to now. Among its most prominent features are a high degree of dependence on bank finance, interlocking ownership among the leading industrial firms, cross-representation on the boards of directors, a low degree of dependence on open capital markets, and (in conjunction with certain provisions in the tax system) a high degree of dependence on retained earnings. This structure has encouraged the development of markets not only for capital but also for technology and managerial competence internally within each enterprise group. While encouraging "intrapreneurship" (entrepreneurship *within* the group, perhaps eventually resulting in a spin-off), it has also tended to discourage entrepreneurship outside these groupings and has therefore had a strong tendency to preserve existing industries and firms to the detriment of emerging ones.

The ownership structures, capital markets, and role of large firms in Finland appear to be quite similar to those in Sweden. The ownership of firms whose stock is traded on the Helsinki stock exchange is relatively concentrated, with a large and increasing share of institutional owners. In 1986, corporations were the largest shareholders in more than 40 % of the companies traded on the stock exchange, while banks (12 %), insurance companies (19 %), and other institutional owners (17 %) together owned nearly half (48 %) of the stock. But the role of the banks is much larger than suggested by these numbers, because similarly to Sweden and Japan, the large commercial banks each have their own financial groups characterized by close ties to the bank, close customer-supplier relations among firms in the group, and mutual stock ownership. (Spolander, 1986.) Similarly to Sweden and Japan, the traditionally high leverage (debt/equity) ratios in Finnish firms means that they are even more dependent on the commercial banks than the banks' share of stock would indicate.

Like in Sweden, the capital markets have traditionally been heavily regulated but have undergone significant deregulation during the 1980s. This is likely to have contributed to a loosening up of the financial groups. The role of venture capital appears to be quite modest, as is the case in virtually all countries except the United States.

---

[2] For an analysis of the development in Japan from *zaibatsu* to *keiretsu* and more recently to *network industrial organization*, see Imai (1989).

While the share of the 30 largest firms in domestic manufacturing employment is somewhat larger in Finland (49 % in 1990) than in Sweden (39 %), their total employment (including employment in foreign operations) represent 80 % in Finland vs. 105 % in Sweden. (Heum and Ylä-Anttila, 1992, 58.)

Thus, it appears that the structures of corporate ownership and control are similar in Finland, Sweden, and Japan but that Finland is not quite as extremely dependent on a few giant domestically based multinational firms as Sweden is.

The Finnish, Swedish, and Japanese pattern may be contrasted with that in the United States. As described by Chandler (1990), one of the distinctive features of the American economy in the latter half of the 19th century was the emergence of a huge domestic market, much larger than any other national market. This development was based on a few major innovations, such as the steamship, the railroad, and the telegraph. But it also depended upon a number of innovations in the financial markets (new information and reporting systems, such as financial and cost accounting, new financing arrangements in the form of bank syndicates, bond markets, and vastly expanded stock markets). The creation of a huge domestic market gave rise to the formation of giant industrial firms (e.g. Standard Oil, U.S. Steel, and DuPont). But along with antitrust legislation, the new institutional arrangements led to a much more dispersed ownership of industrial firms than in most other countries and to much greater reliance on open, external (as distinct from internal) markets for both capital and labor.

7.4 Research and Development

Vuori and Vuorinen (in this volume) provide an excellent overview of the industrial R&D in Finland, both historically and in international comparison. There is not much to add to their presentation, except the following observations. In Table 7.2, data are presented on R&D intensity in various manufacturing industries in Japan, Sweden, and the United States in comparison with that in Finland. The picture that emerges is that while the Finnish R&D intensity is not impressive at the total industry level, it is high or extremely high in the sectors which form the core of Finnish industrial competence. These industries include several engineering industries, namely electrical machinery, electronic equipment and components, certain types of machinery, as well as computers and instruments. However, the areas of particular Finnish strength (the forest-based industries) are characterized by such low R&D intensities that even though Finland spends relatively heavily on R&D in these industries, their over-representation in the Finnish industrial structure actually reduces the overall R&D intensity. On the other hand, Finland also spends heavily on R&D in the relatively R&D-intensive chemical industries. A similar picture emerges for Sweden as well - the high R&D intensity in the motor vehicles, machinery, and steel industries (the latter reflecting the Swedish concentration on

Table 7.2. Research and development expenditures in percent of value added in manufacturing industries in Finland, Japan, Sweden, and the United States, circa 1988

Industry	Finland 1989	Japan 1988	Sweden 1987	USA 1988
Electrical machinery	10.3	10.5	9.7	2.6
Electronic eq. & compon.	26.0	17.6	26.9	24.9
Sub-total electr. group	17.5	14.6	18.5	15.6
Chemicals	9.5	10.0	4.6	5.5
Drugs	24.2	12.9	34.9	15.3
Petroleum refineries	8.8	6.0	0.8	7.7
Sub-total chemical group	11.0	10.4	9.8	7.7
Sub-total aerospace	n.a.	n.a.	n.a.	n.a.
Motor vehicles	4.3	9.5	17.3	n.a.
Ships	4.1	21.9	2.6	n.a.
Other transports	5.1	4.9	8.1	n.a.
Sub-total other transports	4.4	9.9	14.6	7.0
Ferrous metals	2.2	3.9	6.6	0.8
Non-ferrous metals	3.4	8.0	1.0	2.2
Fabricated metal products	2.3	1.2	1.3	1.4
Sub-total basic metals group	2.5	3.0	2.9	1.3
Instruments	14.7	14.6	13.3	7.7
Off. machinery & computers	37.1	n.a.	38.3	n.a.
Machinery NEC	6.9	4.2	11.5	2.5
Sub-total machinery group	9.2	4.3	14.5	n.a.
Food, drink & tobacco	2.0	1.9	1.8	n.a.
Textiles & clothing	0.8	1.3	0.8	n.a.
Rubber & plastics	2.6	3.7	1.9	n.a.
Sub-total chemic. linked group	1.9	2.1	1.7	n.a.
Stone, clay & glass	2.8	3.9	1.5	n.a.
Paper & printing	1.4	0.8	2.4	n.a.
Wood, cork & furniture	0.6	0.6	0.4	n.a.
Other manufacturing	1.5	3.4	8.0	n.a.
Sub-total other manufacturing group	1.4	1.8	1.9	n.a.
Total manufacturing industry	4.7	6.1	7.7	7.3

Sources: OECD, Basic Science and Technology Statistics, 1991
United Nations, Industrial Statistics Yearbook, vol. 1, various issues
Statistics Finland, Koulutus ja tutkimus 1990:24

specialty steel products), is particularly noteworthy - but because of the broader base of the Swedish engineering industry and its larger size compared to Finland, as well as the relatively smaller size of the paper and pulp industry, the total R&D intensity in Swedish industry is quite high.

7.5 Patenting Activity

An examination of patent data reveals the same general picture as that based on R&D data. Figure 7.7 shows the number of patents per million inhabitants in sixteen industrial countries. Switzerland, the United States, Japan, Germany, and Sweden constitute the leading group with more than 100 patents. Finland belongs to the second group of countries (with 50-100 patents); this group also includes Canada, the Netherlands, France, the United Kingdom, and Austria.

In Figure 7.8, data are presented on the number of patents per million inhabitants in selected industries. The patenting "intensity" varies considerably among industries, with the United States and Japan generally showing the highest intensity and Finland and Sweden trailing behind.[3] An interpretation, consistent with the R&D data, is that Swedish and Finnish firms put less emphasis on **R**esearch and more on **D**evelopment than do their counterparts in the U.S. and Japan.

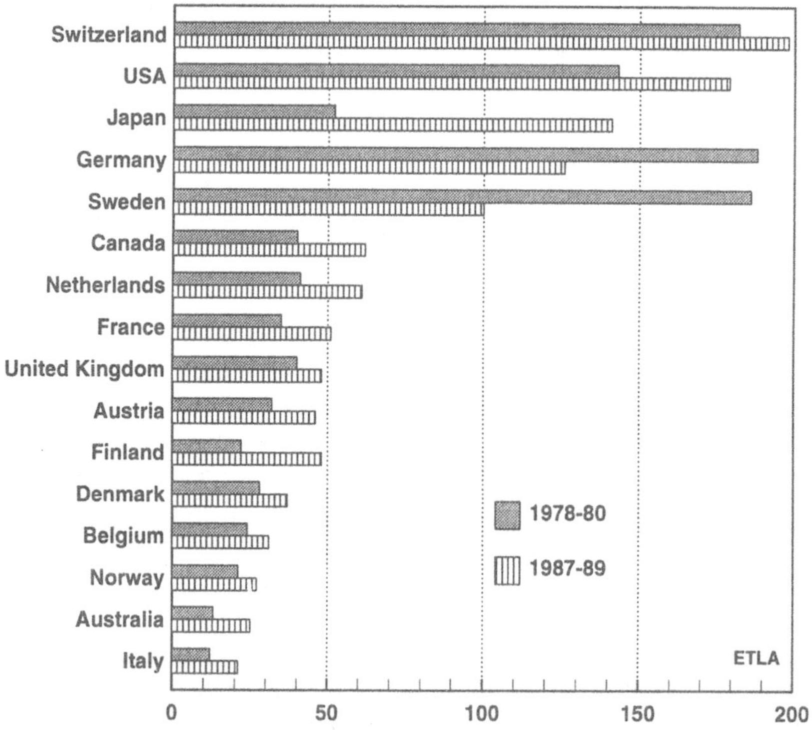

Fig. 7.7. Number of patents per million inhabitants. Source: NUTEK.

[3] The results in Figures 7.7 and 7.8 seem to be in general agreement with the results of G.F. Ray (1988 and 1990).

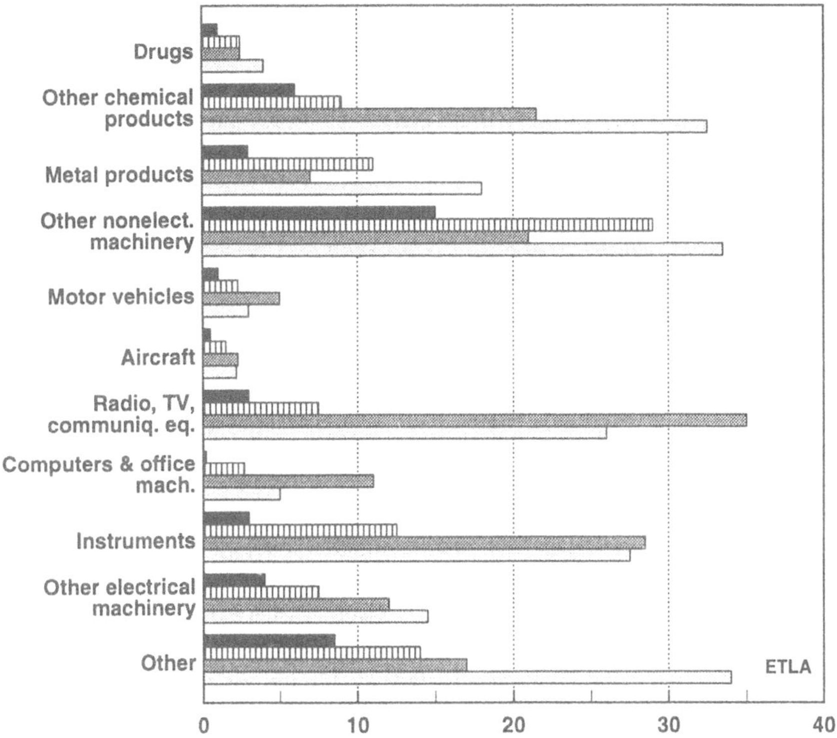

Fig. 7.8. Number of patents per million inhabitants. Source: NUTEK.

7.6 Technological Systems

The data presented in the preceding sections suggest that the environment for technological innovation varies considerably from one industry or area of technology to another. This observation is one of the starting points for a Swedish study, started in 1987 and still going on, of the "technological systems" supporting innovations in various areas of technology.[4]

[4] The project "Sweden's Technological Systems and Future Development Potential" was initiated and is being funded by the National Board for Industrial and Technical Development (NUTEK, formerly STU), with additional funding from the Swedish Council for Planning and Coordination of Research (FRN). A summary of the preliminary findings is presented in Carlsson (1992). A complete list of reports from the project may be obtained from the author.

The importance of *systems* in the analysis of economic growth was pointed out by Freeman (1987), Lundvall (1988), and others, who suggested the concept of a national system of innovation. A national system of innovation may be defined as a "set of distinct institutions which jointly and individually contribute to the development and diffusion of new technologies and which provides the framework within which governments form and implement policies to influence the innovation process" (Metcalfe 1992, p. 82).

Technological systems are similar to national systems of innovation. They have been defined as *network(s) of agents interacting in each specific technology area under a particular institutional infrastructure for the purpose of generating, diffusing, and utilizing technology* (Carlsson and Stankiewicz 1991, p. 111).

While the basic concepts are obviously quite similar (particularly with regard to the systems aspects), technological systems differ from national systems of innovation in three important dimensions. First of all, technological systems are defined by technology rather than national boundaries. They are not necessarily bounded by national borders, although they are certainly influenced by cultural, linguistic, and other circumstances which facilitate or impede contacts among units within the system. Indeed, an important dimension in which technological systems may differ from each other is the degree to which they are international in character.

Secondly, technological systems vary in character and extent from one technology area to another within any given country. For example, the number and characteristics of actors and their interdependence, the institutional infrastructure, the geographic concentration and the degree of internationalization vary among technology areas. A country may be strong in one technology area and weak in another. Thus, Japan appears to be extremely strong in "mechatronics" (the intersection of mechanical engineering and electronics) but not in other manufacturing industries such as chemicals and drugs, nor in distribution and other service industries. (Imai and Yamazaki 1992)

A third difference between technological systems and national systems of innovation is the degree of emphasis on diffusion and utilization as distinct from creation of new technology. As a result, technological systems tend to place more emphasis on the microeconomic (as distinct from macro-oriented public policy) aspects of technology diffusion and utilization. The creation of new technology pushes out the production possibility frontier or opportunity set. But it cannot be simply assumed that just because a technology exists, it is also known and used effectively. Unless the expanded opportunity set is converted into economic activity, i.e., unless it results in entrepreneurial activity, it has no economic impact.

7.6.1 Dimensions of Technological Systems

The dimensions of technological systems have been explored in a Swedish study. The technological systems studied are those supporting factory automation, electronics and computer technology, pharmaceuticals (especially biotechnology) and powder technology.

These four case studies have led to the identification of ten dimensions which may be used to describe technological systems. These dimensions are listed in Table 7.3. For each technological system, the number of pluses and minuses indicates the relative strength in each dimension; this (somewhat subjective) rating will be commented upon in the next section.

Most of the dimensions in Table 7.3 are self-explanatory. The *future potential* of a technological system is largely dependent on where basic underlying technologies are currently located in their life cycle, i.e., their present development phase. *Bridging institutions* refer to arrangements and/or organizations which establish and maintain interaction among various actors in the system. For example, in the area of factory automation a private research institute (IVF), owned jointly by the National Board for Industrial and Technical Development (NUTEK, a government agency) and Mekanförbundet (the engineering industry association), has played such a role.[5] One of its functions is to scan the world for new technology (this scanning is often aided and complemented by the scanning activities of Swedish multinational firms), evaluate it and rapidly disseminate information on it through

Table 7.3. Dimensions and characteristics of technological systems

	Factory Automation	Electronics	Pharma-ceuticals	Powder technology
Present development phase	Mature	Rapid dev.	Rapid dev.	Heterogeneous
Future potential	+	+++	+++	++
Buyer competence	+++	++	+++	+
Buyer-supplier collaboration	++	++	+++	+
Supplier competence	++	++	++	++
Industrial R&D	+++	+	+++	++
Academic infrastructure	+	+	+++	++
Government policy	+	+	++	++
Bridging institutions	+++	+	+++	++
Holes/weaknesses	-	- -		-
Compensating mechanisms	++	+	+++	+++

Source: Carlsson (1992b).

[5] IVF was set up in 1964 to conduct scanning, monitoring, adaptation, and diffusion of production technology in the engineering industry, as well as contract research and testing for individual firms. All of these organizations are cross-represented on numerous committees and task forces, where small as well as large firms also take part. Academic institutions are represented, primarily in NUTEK and IVF. Thus, IVF and NUTEK provide links between academia and business, while Mekanförbundet bridges the gap between government and industry.

the network of which it is a part, thus fostering a high level of technical awareness at all levels of Swedish industry which is central to explaining the rapid diffusion of technology. In addition, bridging institutions such as this help to accumulate and integrate the results of innovative activities which otherwise tend to be highly firm-specific and make them useful and available to other firms as well.

The role of these organizations is not only to disseminate know-how but also to provide a *compensating mechanism* for *weaknesses* and lack of domestic capability ("holes") within other parts of the technological system. In some technological systems such weaknesses appear in the form of absence of domestic suppliers of key technologies, lack of buyer competence, lack of research capability in the relevant part of the academic system, etc. In Sweden, a particularly important compensating mechanism is the global monitoring and diffusion of technology via multinational firms. This reflects both the relatively small size of the country and the large role played by Sweden-based multinationals in the economy.

7.6.2 Characteristics of Technological Systems: Results of Four Swedish Case Studies

Due to space limitations it is impossible here to discuss the key characteristics of each technological system.[6] Only a brief summary of the principal findings is presented.

There are several important features which all of these technological systems have in common. First, our study indicates that it may not be sufficient for sustainable economic growth to have only one or a few competent actors in a given system. There has to be a *variety of actors,* each with *specific* (sometimes unique) *competence.* They must also act together, in *clusters or networks.* Given the risk taking necessarily involved in new activities, having a supportive network reduces the risk of any given venture by providing timely information: early identification and quick feedback increase the possibilities of taking corrective action whenever necessary.

---

[6] The four cases were originally selected for the following reasons. Factory automation is a broad area of technology where Sweden is very strong internationally. It has a long history in Sweden and can therefore be expected to exhibit all the characteristics of a fully developed technological system. Electronics and computers is an even broader technical area involving some of the same actors as factory automation but where the Swedish capability is limited and the technological system much less developed. Pharmaceuticals were selected because they represent an entirely different type of technology. Initially, the primary focus was on biotechnology, but it was soon discovered that there is almost no use of biotechnology in Sweden outside the pharmaceutical industry. Powder technology, finally, was selected because it represents a much more embryonic field, offering possibilities of observing an emerging technological system.

Second, the crucial role of *competent buyers* is illustrated in each of the investigated systems. These buyers are often industrial firms but may also be government agencies. *Close collaboration*, often on a continuing and long-term basis, *between buyers and suppliers* seems to be highly beneficial, if not essential.

Third, in spite of the trend toward increased internationalization of all kinds of economic and technical activity, our study suggests that strong *local or domestic technological systems continue to be important*, even in small countries, and even though their links to systems in other countries may also be important, particularly via multinational firms. For example, the factory automation case study shows that while leading users may not be directly dependent on local or domestic suppliers of automation technology, the majority of less advanced users are still highly dependent on competent domestic suppliers. (Carlsson and Jacobsson 1991.) In the absence of such suppliers of technology, the leading users may find themselves without domestic suppliers of parts and components, making their continued domestic production less competitive and increasing the attractiveness of moving production elsewhere, thus further weakening the domestic industrial base. In other words, the technological system may be instrumental in keeping domestic firms firmly rooted at home while at the same time attracting foreign firms.

While there are thus some key features which these four technological systems have in common, there are also dimensions in which they differ from each other. First, they differ with respect to future development potential (depending partly on the particular stage of development of the underlying technology in each case) and the extent to which market opportunities have been exploited through the creation of strong networks, infrastructure, and competence on the part of various economic agents.

Secondly, they vary with respect to where within the system new knowledge is being generated. In mechanical engineering, technological change appears to be relatively slow, "routinized," and codified, meaning both that new technical developments occur to a relatively great extent in universities and that they then become relatively easily accessible to users. In the area of pharmaceuticals, the knowledge base is expanding much faster; technological change tends to take place outside the universities or in joint ventures between pharmaceutical firms and universities. In electronics, the knowledge base is expanding even more rapidly, with the result that academia rarely has the competence to offer advanced knowledge. Instead, universities primarily play the role of supplying basic engineering training, while most research and development takes place in business firms. Frontier knowledge is typically tacit, since receiver competence is often lacking (Eliasson 1989, 40). As a result, strong domestic academic competence is important in some technological systems but much less so in other systems.

In the area of factory automation, which is relatively mature technically, the future potential for generating new economic growth is limited in spite of a high user competence and a well-functioning network. Nevertheless, if the now mature technology is diffused widely to new user groups, it may increase the competitiveness and therefore the growth potential of existing industries. An important question is whether the existing compensating mechanisms in the system are strong enough to make up for the holes which are beginning to emerge on the supplier side; several key Swedish machine tool suppliers are currently facing bankruptcy.

In the pharmaceutical area, a high level of competence of both users and suppliers and a relatively extensive network, in combination with continued rapid technological change, mean that the future development potential is extraordinarily great. The possibilities of taking advantage of the market opportunities seem to be good, even though there are some doubts as to whether viable new firms can be created both in the pharmaceutical industry and in neighboring areas of application of biotechnology. The historical record of firm formation is not strong in either of these areas in Sweden.

In the electronics area, the pace of technological change is also very high and the future potential great. The competence level of users and suppliers is very high but spotty; the question is whether the existing competence has sufficient coverage or critical mass to enable the networks to function well. The infrastructure here is not extensive.

Finally, as regards powder technology, the competence level is high, particularly among suppliers on the powder metallurgy side. In ceramics, by contrast, the infrastructure is relatively extensive, especially in the universities, but the industrial competence is still fairly undeveloped. It is therefore difficult to judge the future potential for the technology area as a whole.

As a result of the differences just mentioned, the role of the infrastructure varies from one case to another. In Sweden, the academic part of the infrastructure seems to play a crucial role in pharmaceuticals and powder ceramics, while it does not play a leading role in relatively well-established technology areas such as factory automation and powder metallurgy; it may actually have formed an impediment in the electronics and computer technology area. The rest of the infrastructure (research institutes, branch organizations, government agencies, etc.) also plays varying roles, ranging from highly significant (in the form of bridging institutions) in the factory automation area to a very modest role in the pharmaceutical area.

7.6.3 International Comparison of Technological Systems

The implication of the preceding section is that it may be necessary to go beyond national systems of innovation in the analysis of a country's industrial performance and to examine technological systems instead. The Swedish study suggests that the analysis of technological systems requires fairly detailed and specific information which is not generally available but has to be collected as needed. Therefore, it is not possible within the confines of this paper to make a full-fledged international comparison. Instead, what will be attempted here is only a broad outline of what such a comparison might look like in one particular case, namely the technological system supporting factory automation. Given the pervasive role of factory automation in the engineering industries, it may not be too unreasonable to make inferences to the broader industry. The comparison is based on detailed knowledge concerning the Swedish system and some reflections on bits of evidence concerning the United States, Japan, and Finland.

United States. Even prior to World War II, the U.S. had the strongest technological base of any country, especially in engineering. The U.S. gained greatly in economic power in connection with World War II (as was the case earlier in connection with World War I); the war effort itself led to a massive re-building of American industry, particularly in metalworking (engineering), taking advantage of prewar technological trends; and the physical destruction of its main competitors gave the U.S. undisputed leadership in many areas of industrial activity. This leadership was further strengthened through the vast expansion of higher education which took place in the United States immediately after the war, partly as a result of federal government funding of the education of millions of U.S. soldiers returning to civilian life. In addition, the large-scale defense-related R&D spending which began during the war continued in connection with the Cold War and culminated in the space effort. Thus, the U.S. leadership in industrial R&D is a postwar phenomenon and is connected to defense; in terms of civilian R&D, both Germany and Japan have outspent the United States (as a percentage of GNP) since the early 1970s. (U.S. General Accounting Office 1992, 27-28.)

The impact of the high R&D spending (both military and civilian) on the U.S. technology base and civilian economy appears to have been enormous and largely beneficial. But after 1970 there seems to have been increasing divergence between military and civilian requirements. Thus, there seem to have been fewer areas of civilian production benefiting from military research, and there even seems to be an increasing number of cases in which new products are transferred from the civilian to the military sector rather than vice versa.

One important beneficiary of military R&D has been the machine tool industry; the relative decline in military spending on R&D and on procurement, in combination with increasing divergence between military and civilian technical requirements has led to the virtual demise of the U.S. machine tool industry; civilian American advanced users of machine tools are not numerous and competent enough to sustain the domestic machine tool industry at its historical level; this means increasing difficulties for second echelon firms (i.e., the bulk of metalworking industry) to obtain the most up-to-date technology and hence to compete in global markets.[7] Meanwhile, Japanese and German machine tool firms have developed flexible technology for cheap production of a great variety of parts produced in small batches, thereby undermining mass production of standardized parts produced in large volume.

---

[7] For a more extensive analysis of the development of the U.S. machine tool industry, see Carlsson (1989). For an analysis of the impact of the decline of the U.S. machine tool industry on the competitiveness of U.S. engineering industries, see Carlsson & Taymaz (1993) and Carlsson & Jacobsson (1991).

Thus, the impression one gets is that since World War II, the U.S. military has been the major force shaping technological innovation in the engineering industries in the United States. It has played this role by 1) issuing clear technical challenges and being a competent buyer, and 2) directly and indirectly shaping the technological systems supporting military production. This system seems to have worked well until the early 1970s. But then technological divergence between military and civilian requirements and rapid progress in non-defense areas elsewhere began to shift technological advantage away from the United States.[8] This tendency has been aggravated by the low appreciation of technological interdependence and of the public value of coordinated standards characterizing U.S. industry and public policy. Institutional forums for cooperative activity are weak not only because of free market ideology and antitrust policy and enforcement but also because of the relatively weak and inconsistent role played by U.S. trade associations and professional societies in technology development and commercialization. (Aram, Lynn and Reddy 1992.)

Japan. The U.S. development is almost opposite to that in Japan, where the role of the military has been severely circumscribed during the postwar period and where there are strong cooperative tendencies through the *keiretsu* system, government-business interaction, and weak antitrust laws.

The evolution of the Japanese industrial system in the postwar period has been analyzed by Imai (1989) and Imai and Yamazaki (1992). These analyses emphasize the role of group entrepreneurship, first in the form of *keiretsu* and later transformed into *network industrial organization,* involving close user-producer interaction, with a key role of large firms in initiating these linkages, and with banks playing an important role as "investment bankers" rather than "commercial bankers," i.e., by establishing long-term relations with firms through part ownership, part long-term lending. The user-producer linkages became more complex and at the same time more flexible as the needs shifted from mechanical processing technology in the mature mechanical industries to broader coordinated *systems* for processing, assembling, and testing of electronic products. Thus, technological and organizational changes took place simultaneously, with banks and trading companies as intermediaries, and with job rotation and in-house education on electronics and information technologies ensuring the acquisition of necessary labor skills. Large firms have played a dual role as both producers and users of flexible automation systems, the networks having ensured rapid introduction of automation, not only in large but also in small firms. Imai and Yamazaki also note the importance of the existence of *several* firms, both large and not so large, not just one

---

[8] However, several of the areas of technology in which the United States is still considered to be leading, namely space and aviation, biotechnology, medical and pharmaceutical products, artificial intelligence, new industrial materials, workstations, supercomputers, microprocessors, and personal computers, had their origins in defense or defense-related activities. (Rosenblatt 1991.)

or two giants, for the generation, diffusion, and utilization of new technology. They point out the role of big firms in forming networks for technology diffusion, reducing transaction costs associated with differentiated and complex capital goods, and creating a *critical mass* of demands in the early stage of diffusion.

In short, the picture that emerges is very similar to the technology system for factory automation (and probably for the engineering industry more generally) in Sweden. Imai and Yamazaki also note that while Japan has been highly successful in "mechatronics," the Japanese performance has been much less successful in other areas of technology, such as chemicals, distribution, and several other service sectors. This confirms the finding in the Swedish study that technological systems have different characteristics in different areas of technology and also suggests that the determinants of the "Japanese miracle" may be much less uniquely Japanese than commonly perceived and more defined by the particular requirements of certain technologies.

Finland. Without a detailed study of the technological system supporting the engineering industries in Finland, it is not possible to make any definitive assessment; instead, it seems more useful to indicate the main questions that must be answered in order to get a more complete picture.

As a result of the relatively late industrialization in Finland and the high degree of dependence on forest raw materials, the engineering sector is still small in comparison with that in the other countries in this study, in spite of its rapid growth in recent decades. Its output also appears to be relatively strongly oriented towards the forest-based industries. Thus, the development bloc based on the harvesting and processing of forest raw materials continues to play an important role in Finnish industry. Therefore, an important question for the future is what the linkages are between this development bloc and the larger one (actual and potential) encompassing the whole set of engineering industries.

In order to make a more complete assessment of the technological system for factory automation in Finland, the following set of questions has to be answered. What is the nature of the academic and institutional infrastructure? What is the role of domestic suppliers of machine tools and other automation technology? Given the historically close links between Finland and Sweden, can Swedish suppliers be regarded as substitutes for or complements to domestic ones (thus to some extent reducing the need for domestic suppliers)? To what extent is the Finnish technological system in this area integrated with international systems? What is the role of leading users of automation technology? Are they sufficiently competent and numerous to sustain a domestic supplier industry? To what extent do the smallness of the country and the existence of bridging institutions similar to those in Sweden provide strength and coherence to the technological system? And what is the role of government policy in this area?

7.7 Conclusion

What, then, can we conclude from this overview and international comparison? Let us first summarize what we have learned.

Sweden appears to have fairly strong technological systems in the areas investigated thus far and performs well in terms of both R&D and results of R&D (as measured by patents). In spite of this, Sweden has had a dismal growth record during the last two decades and has lost international market share. In my view, this is due to problems in the domestic economic environment having to do more with institutions than technology: the welfare state with high taxes and poor incentives, too much of the economy being guided by non-market principles, corporativistic structures, and lack of EC membership. These have caused a lack of entrepreneurship (and hence too much dependence on a handful of MNFs) and has probably caused Sweden-based MNFs to expand more abroad than they might have otherwise. As a result, the market share of Sweden-based firms and Swedish exports combined has held up much better than Swedish exports alone. For similar reasons, Sweden has not been an attractive environment for inbound direct foreign investment. Even though the technological systems appear strong, they may not be large and robust enough to support the global (not just domestic) activities of Sweden-based multinational firms and to attract foreign-based multinationals to Sweden.

Finland continues to be highly dependent on its forest-based industries where it is strongly competitive and appears to have a robust technological system. It is also strong in some segments of the engineering industries, but because of the small number of domestic firms (i.e. lack of critical mass), Finland is becoming more dependent on collaboration with foreign firms. While Finnish industry invests heavily in R&D in the segments where it is strong (forest-based industries and certain segments of the engineering industries - particularly those supplying equipment to the forest-based industries), thus contributing to its continued competitiveness in these sectors, its overall R&D spending level is not impressive in international comparison because of the structure of output. The patenting activity also seems modest. Similarly to Sweden, given the country's location remote from major European markets, high production costs, and lack of EC membership, Finland, too, may need an "oversize" technology base in order to constitute an attractive base for both domestic and foreign firms. But Finland appears to have done much better than Sweden in creating a favorable environment for entrepreneurship in the form of new business formation.

Japan's strong growth record until recently appears to be based on well-functioning technological systems in the area of mechatronics, the country having shifted its comparative advantage from labor-intensive industries in the early post-war period first into steel, then automobiles and, more recently, electronics. The technological systems supporting innovation in the engineering sector seem to be remarkably similar to the system for factory automation studied in detail in

Sweden. But Japan's performance is significantly less impressive in other industrial sectors and in many service industries. This suggests that further study of technological systems in other sectors would be rewarding in gauging the future growth potential of the Japanese economy.

The United States had strong, well-functioning technological systems in the engineering sector in the early postwar decades, largely as a result of the "military-industrial complex." As that complex began to crumble (for a variety of reasons), U.S. economic performance declined, even though the country remained strong in innovation and higher education. The areas of technology in which the U.S. continues to lead were originally based on military requirements but are rapidly shifting to new areas of application. The U.S. also appears strong in non-engineering technologies (esp. software and biotechnology) as well as in many rapidly growing service sectors.

At the very least, this essay has suggested that technological systems may be helpful in understanding economic growth - but that they are also sometimes overshadowed by other (mostly institutional) factors. It has also indicated the need to study specific technological systems and make more in-depth international comparisons. Given the differences among the systems supporting innovation in various technological areas, it seems to make more sense to carry out such studies at the technological system level rather than at the level of the national system of innovation as a whole.

As previous studies have shown (Carlsson 1992b and 1992c, Carlsson & Jacobsson 1992), there are numerous implications of this type of analysis for public policy as well as business management. But in line with the previous discussion, the implications are more specific (to each technological system) than general. Therefore, the policy implications for Finland need to have much better and more precise empirical underpinnings than have been provided in this paper. They warrant a separate and careful study.

References

Aram, J.D., L.L. Lynn and N.M. Reddy, N.M. (1992), "Institutional Relationships and Technology Commercialization: Limitations of Market-Based Policy." Research Policy, vol. 21, No. 3, 409-421.

Blomström, M. and R. Lipsey (1989), "The Export Performance of U.S. and Swedish Multinationals." Review of Income and Wealth, Series 35, No. 3 (September), 245-264.

Carlsson, B. (1989), "Small-Scale Industry at a Crossroads: U.S. Machine Tools in Global Perspective." Small Business Economics, vol. 1, No. 4, 245-261.

Carlsson, B. (1990), "Technology and Competitiveness: The Micro-Macro Links - A Comparison between the United States and Sweden." Paper presented to the OECD Conference on Technology and Competitiveness, Paris, June 24-27.

Carlsson, B. (1992a), "The Rise of Small Business: Causes and Consequences", in: W.J. Adams (ed.), Singular Europe: Economy and Polity of the European Community after 1992. Ann Arbor, MI: University of Michigan Press.

Carlsson, B. (1992b), "Technological Systems and Economic Development Potential: Four Swedish Case Studies." Paper presented to the International Joseph A. Schumpeter Society Conference, Kyoto, August 19-22.

Carlsson, B. (1992c), "Small Country Squeeze and Technological Systems: Implications for Management." Case Western Reserve University, Department of Economics working paper 92.07.

Carlsson, B. (1992d), "Industrial Dynamics and the Role of Small Plants in Swedish Manufacturing Industry, 1968-1988." Case Western Reserve University, Department of Economics working paper 92.19.

Carlsson, B. and S. Jacobsson (1991), "What Makes the Automation Industry Strategic?" Economics of Innovation and New Technology, vol. 1, No. 4, 257-269.

Carlsson, B. and S. Jacobsson (1992), "Technological Systems and Economic Policy: The Diffusion of Factory Automation in Sweden." Working paper, Case Western Reserve University and Chalmers University of Technology.

Carlsson, B. and R. Stankiewicz (1991), "On the Nature, Function, and Composition of Technological Systems." Journal of Evolutionary Economies, Vol. 1, No. 2, 93-118.

Carlsson, B. and E. Taymaz (1993), "Technological Capabilities and International Competitiveness in the Engineering Industries." Review of Industrial Organization (forthcoming).

Carlsson, B., E. Dahmén, A. Grufman, M. Josefsson and J. Örtengren (1979), Teknik och industristruktur - 70-talets ekonomiska kris i historisk belysning (Technology and Industrial Structure - The Economic Crisis of the 70's in Historical Perspective). Stockholm: Industriens Utredningsinstitut and Royal Swedish Academy of Engineering Sciences.

Chandler, Jr., A. (1990), Scale and Scope: The Dynamics of Industrial Capitalism. Cambridge, MA: Harvard University Press.

Edquist, C. and B.-Å. Lundvall (1989), "Comparing Small Nordic Systems of Innovation," paper presented at the Maastricht seminar on National Systems Supporting Technical Progress, November.

Eliasson, G. (1989), "The Dynamics of Supply and Economic Growth - How Industrial Knowledge Accumulation Drives a Path-Dependent Economic Process," in: Carlsson, B. (ed.), Industrial Dynamics: Technological, Organizational, and Structural Changes in Industries and Firms. Boston and Dordrecht: Kluwer Academic Publishers.

Encarnation, D.J. (1992), Rivals beyond Trade: America versus Japan in Global Competition. Ithaca: Cornell University Press.

Freeman, C. (1987), "National Systems of Innovation: the Case of Japan," Ch. 2 in: Technology Policy and Economic Performance: Lessons from Japan. London: Pinter Publishers.

Heum, P. and P. Ylä-Anttila (1992), "Firm Dynamics in a Nordic Perspective: Large Corporations and Industrial Transformation." ETLA (The Research Institute of the Finnish Economy), Discussion Paper no. 423, December (in cooperation with P. Braunerhjelm and S. Thomsen).

Imai, K. (1989), "Evolution of Japan's Corporate and Industrial Networks," in: Carlsson, B. (ed.), Industrial Dynamics: Technological, Organizational, and Structural Changes in Industries and Firms. Boston and Dordrecht: Kluwer Academic Publishers.

Imai, K. and A. Yamazaki (1992), "Dynamics of the Japanese Industrial System from a Schumpeterian Perspective." Paper presented to the International J.A. Schumpeter Society conference, Kyoto, August 19-22.

Kajaste, I., S. Parviainen and P. Ylä-Anttila (1992), "Foreign Direct Investment of Finland in the 1990s," VATT (Government Institute for Economic Research), Discussion Paper No. 28.

Lundvall, B.-Å. (1988), "Innovation as an Interactive Process: From User-Producer Interaction to the National System of Innovation", in: G. Dosi et al. (eds.), Technical Change and Economic Theory. London: Pinter Publishers.

Metcalfe, J.S. (1992), "The Economic Foundations of Technology Policy: Equilibrium and Evolutionary Perspectives." Mimeo., (first draft), November.

Ray, G.F. (1988), "Finnish Patenting Activity." ETLA (The Research Institute of the Finnish Economy), Discussion Paper no. 263, June.

Ray, G.F. (1990), "More on Finnish Patenting Activity," ETLA (The Research Institute of the Finnish Economy), Discussion Paper no. 331, July.

Rosenblatt, A. (1991), "Who's Ahead in Hi-Tech?". IEEE Spectrum, April, 21-27.

Spolander, P. (1986), "Kapitalmarknader och ägarförhållanden i Finlands näringsliv" (Capital Markets and Ownership Structures in the Finnish Economy). ETLA (The Research Institute of the Finnish Economy), Discussion Paper no. 225, December (in Swedish).

Swedenborg, B., G. Johansson-Grahn and M. Kinnwall (1988), Den svenska industrins utlandsinvesteringar, 1960-1986 (Foreign Investment by Swedish Industry, 1960-1986). Stockholm: Industriens Utredningsinstitut.

United States General Accounting Office (1992), High-Technology Competitiveness: Trends in U.S. and Foreign Performance. GAO/NSIAD-92-236. Washington, D.C.

8 Characteristics of Technology Policy in Finland

Tarmo Lemola
Group for Technology Studies, Technical Research Centre of Finland, Espoo, Finland

8.1 Introduction

Technology policy has become an increasingly important field of government activity in all the industrialized countries. This phenomenon is by no means as new as is often imagined. Neither has the tempo of change with regard to technology policy necessarily been as rapid as its ardent proponents have assumed. Nevertheless, it is certainly true that technology policy has been one of the most dynamic instruments of the public sector.

In the initial stage of the progress of nation states, government measures were primarily aimed at establishing and developing of the basic structures of a scientific and technological system, e.g. universities and research institutions (Löppönen and Noro 1975, 52). After the Second World War, the technology policies of the superpowers were oriented towards the development of military and nuclear power technologies. The scope of technology policy was later expanded to encompass space technology (Freeman 1982).

In the late 1960s, the development needs of industry began to assume a more important position in the priorities of technology policy (OECD 1980, Rothwell and Zegveld 1981). This stage also coincided with the shaping of technology policy in its present-day form in small countries. The creation of new corporate activities and the support of innovation in existing firms became the key objectives of technology policy.

The fundamental goals and points of departure for technology policy did not change essentially during the 1970s and 1980s. The volumes of technology policy action certainly grew and the range of instruments became more diverse. The most important new challenge facing national technology policy has been the expansion of international cooperation and the role of transnational organizations (Sigurdson 1989).

The technology policies of different countries have converged heavily as a result of the interaction of increased cooperation and stiffer international competition. Japan has played a very important and conspicuous role in this convergence process. Japan's influence has also been apparent in Finland, but Sweden still holds more sway over the Finns. The OECD was also an important source of ideas and models for the early designers of Finnish science and technology policy.

In recent years technology policy has succeeded in resisting the pressures of deregulation better than any other form of government intervention. This has also been the case in those countries which have stuck more doggedly to monetarist economic policies, e.g. the United Kingdom and the United States. With respect to the future of technology policy, it is also worth noting that, even if European nation states will have to dismantle the systems they have built to support various economic sectors, the support of corporate R&D work will remain outside the scope of such measures.

Technology policy has been the subject of considerable research work in recent years. This work has been approached from different angles, including that of a national innovation system, by examining technology policy as part of a broader national system (Lundvall 1992). On the other hand, individual instruments of technology policy have also been evaluated (Ormala 1987). Information on the impetus of technology policy and on the effects of various measures has increased, but really convincing answers to the classically posed question "Does technology policy matter?" (Ergas 1986) have been few and far between.

8.2 The Development Phases of Finnish Technology Policy

In Finland the development of technology policy has followed the general international trend. Sweden and the OECD have been the sources of the models and ideas adopted in Finland (Luukkonen-Gronow 1975 and 1977). The construction and development of the machinery for technology policy occurred in Finland later than in the major and more advanced OECD countries. However, the rapid pace of development in Finland since the latter half of the 1960s compensated for this late start. The national features of Finnish technology policy are based on the absolute and relative shortage of resources, the relative insignificance of military research and other fields of big science, and the small number of research-intensive firms and industrial branches producing and using advanced technology.

The first educational institutions serving the needs of productive industry in Finland were established around the turn of the century. The Geological Research Institute and the Agricultural Research Centre, founded in 1885 and 1889 respectively, were the first state research establishments in Finland. The Forest Research Institute, which was clearly linked to production, i.e. the supply of raw materials to the forest industries, was set up in 1917. The Finnish Pulp and Paper Research Institute, a joint-stock research enterprise owned by the forest industries, had been founded a couple of years earlier. The old Polytechnic became the Helsinki University of Technology in 1908, and in 1921 the Faculty of Chemistry and Technology was set up at the Åbo Academy. The next important step towards developing technical research serving the needs of industrial production was taken during the war in 1942, when the Technical Research Centre of Finland (VTT) was founded.

In the 1950s the technical sciences and their economic importance became the subject of more focused debate (Falk, Jaakkola & Viikari 1975; 97 - 98). Activities associated with the peaceful use of atomic energy served as an important catalyst in this debate (Laurila 1982). Nevertheless, it was not until the mid-1960s that Finland set about constructing the machinery for a genuine technology policy.

A number of factors turned the spotlight on the need for broad-based development of Finnish technology policy. Intensified internationalisation and the liberalisation of trade, which first affected Finland in the late 1950s, placed enormous new strains on Finland's one-sided production structure, and her level of technology, which was low compared with her main competitors. The doctrine of economic growth, which in the early 1960s also gained a foothold in Finland, significantly promoted industrial expansion and increased the role and tasks of the state in promoting and supporting innovation in the corporate sector (Kosonen 1981, 74-77).

Three important changes occurred in Finnish technology policy at the turn of the 1970s (Lemola & Lovio 1984):

- The state began to support corporate R&D work directly in the form of product development loans and subsidies.
- The basic operating conditions and requirements of the universities and research institutions were improved. At the same time, the activities of these establishments were reoriented so as to better serve the needs and demands of industry. Particular attention was focused on the VTT.
- The funding of goal-oriented technical research was started by seeking to increase and improve research work which was of national importance and aimed at promoting the longer term development of industry.

These three initiatives can be regarded as the pillars on which Finland's current technology policy rests.

After the machinery for Finnish technology policy was constructed at the beginning of the 1970s, the remainder of the decade was dedicated to the quantitative development of the new system. The main instruments applied during this period - product development and goal-oriented research appropriations and the VTT - grew much faster than the average for state research expenditures, as can be seen in Table 8.1. The same growth rate also continued in the 1980s. The most important of the changes in the mutual relationships between these instruments was the enhanced role of product development and goal-oriented research appropriations.[1]

[1] The information is based on the annual reports of TEKES and the VTT, and on statistical data, at the disposal of the author, on appropriation decisions of the Ministry of Trade and Industry during 1970 - 1982.

Table 8.1. The most important state appropriations intended for technological development, 1970 - 1990 (1)

	Expenditure at 1990 prices, FIM 1000				
	1970	1975	1980	1985	1990
TEKES/industrial R&D	35 838	125 424	182 154	311 610	475 000
TEKES/applied technical research	5 082	16 474	30 434	100 776	226 000
VTT	145 179	296 548	350 130	616 241	769 385
SITRA	52 622	61 316	73 194	75 052	45 600
Total government R&D financing	1 134 865	1 706 227	2 126 000	3 013 998	4 159 000

	Average annual growth rate, %			
	1970-75	1975-80	1980-85	1985-90
TEKES/industrial R&D	28.5	7.7	11.3	8.7
TEKES/applied technical research	34.2	13.1	27.0	17.5
VTT	15.3	3.4	12.0	4.6
SITRA	3.2	3.5	1.0	-9.4
Total government R&D financing	8.4	4.6	7.3	6.7

(1) See footnote 1.
TEKES = Technology Development Centre.
VTT = Technical Research Centre of Finland.
SITRA = Finnish National Fund for Research and Development.

In the latter half of 1970s, it became apparent that increasing the amount of appropriations as such would not be sufficient; in addition, attention would have to be paid to the allocation and use of appropriations. In fact, the lack of goal-orientation began to form a serious obstacle to the continuing quantitative growth of appropriations. At the end of 1980, it was proposed that major projects should be initiated in the key areas of technological development. Subsequently, the proposal led to the introduction of a technology programme procedure.

In order to increase goal-orientation, it was necessary to create an organisational framework. In 1983, the preparatory work resulted in the establishment of TEKES, the Technology Development Centre of Finland. TEKES'es first and most urgent task was the implementation of national technology programmes.

The decisions to set up TEKES and introduce the technology programme procedure were both obviously correct, but they were not particularly original measures.

TEKES was modelled on an organisation established for the same purpose in Sweden in the late 1960s. Models for the technology programme procedure were found not only in Sweden but also in many Western European countries and, of course, in Japan.

In the mid-1980s, the internationalisation of R&D became the latest phase of technology policy. Its most dynamic components have been EUREKA, EC research programmes and space cooperation. The pace of development is well described by the fact that TEKES expended around FIM 10 million on funding for international cooperation in 1985, whereas the corresponding amount was FIM 145 million in 1991. A similar trend, which is partly attributable to the funding provided by TEKES, is also apparent in the VTT and Finland's universities of technology.

8.3 Foundations of Technology Policy

If a technology policy strategy means a consistent plan or programme, based on comprehensive assessments and visions of the development needs and possibilities of technology and defining the technology policy goals and the means by which they are to be achieved, then it must be admitted that Finland does not have, and never has had, an explicit technology policy strategy. Finnish technology policy has been pursued very pragmatically on the basis of knowledge and views of the various parties involved.

One point of departure for technology policy since the 1960s has been a belief in the role of technological competitiveness. Another point has been the view that R&D work, education and training play key roles in improving technological competitiveness. In practice, however, these roles have often been left in the shadow of R&D work. In promoting this work, there has been a prevalent tendency to give priority to the R&D activities of firms.

In 1974 the YTK Committee (1974,15) expressed the matter as follows: "Maintaining the competitiveness of Finnish products in both the domestic and export markets is particularly important for the overall wellbeing of the economy. Moreover, satisfying the needs of society and improving living standards in the long term requires increased output and economic growth. This can only be achieved by bringing the country's production structure in line with both demand and the available productive resources. R&D in the corporate sector can promote the necessary structural changes and thus create the preconditions for continuing economic growth."

According to this chain of conclusions, increasing the resources made available for R&D work has played a prominent role in Finnish technology policy. The demands for increased resources have been based not only on a belief in the importance of know-how and expertise but also on an awareness that Finland has invested much less in R&D than her main competitors, and on information about the measures implemented in other countries. To put it in a nutshell, these two arguments - lagging behind competitors and the measures implemented in other countries - have been the main driving forces of Finnish technology policy.

In retrospect, Finnish technology policy has been largely pursued along the lines of the market failure theory, even though the technology policy-makers have not necessarily been aware of these theoretical starting points.[2] Supporting corporate research and development with loans and subsidies has been one of the most important instruments of technology policy. The aim of this assistance has been to support and augment the corporate sector's own efforts and to compensate for problems arising from the small size of firms. The technology programmes have been used to initiate activities beyond the resources and risk-carrying capacity of firms.

8.4 Quantitative Possibilities of Technology Policy

The effectiveness and successfulness of technology policy largely depend on how well R&D subjects are selected and how well measures are planned and implemented. Even if we give priority to qualitative factors, the quantitative framework should also be borne in mind, if for no other reason than to prevent technology policy from being subjected to unrealistic demands or fears.

The corporate sector's input to R&D increased in Finland significantly in the 1980s. During the ten-year period from 1979 to 1989, the input almost tripled in terms of research expenditures and doubled in terms of research working years. Furthermore, whereas in 1979 the number of licentiates and doctorates engaged in R&D in the corporate sector was 412, the corresponding number 10 years later was 718.

Also, the research intensity of the corporate sector, i.e. R&D expenditures in relation to the value added of output, developed very favourably compared with the preceding decade, as can be seen in Figure 8.1. The research intensity rose from 1.7% in 1979 to 4.3% in 1989. Both the growth in R&D input and intensity of the Finnish corporate sector were faster than in the OECD countries on average.

---

[2] The YTK Committee (1974, 13) expressed this matter as follows: "Research input determined on the basis of decisions made by firms is not always adequate from the standpoint of social goals. R&D may remain at too low a level in those fields where the economic opportunities required by corporate activity are not sufficiently evident ... Government measures should be employed to ensure that the prospects of domestic firms to maintain the research input required by competitiveness are not worse than those of foreign firms."

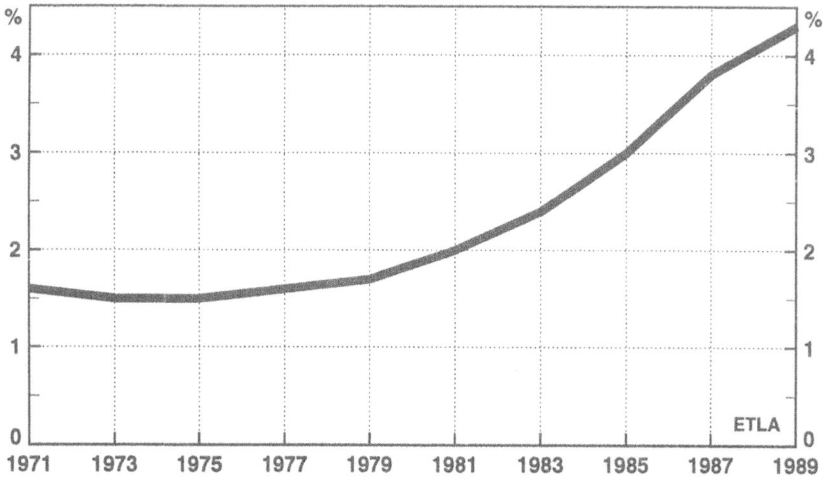

Fig. 8.1. R&D expenditure of the corporate sector in relation to the value added of industrial production in Finland, 1971 - 1989. Source: Statistics Finland.

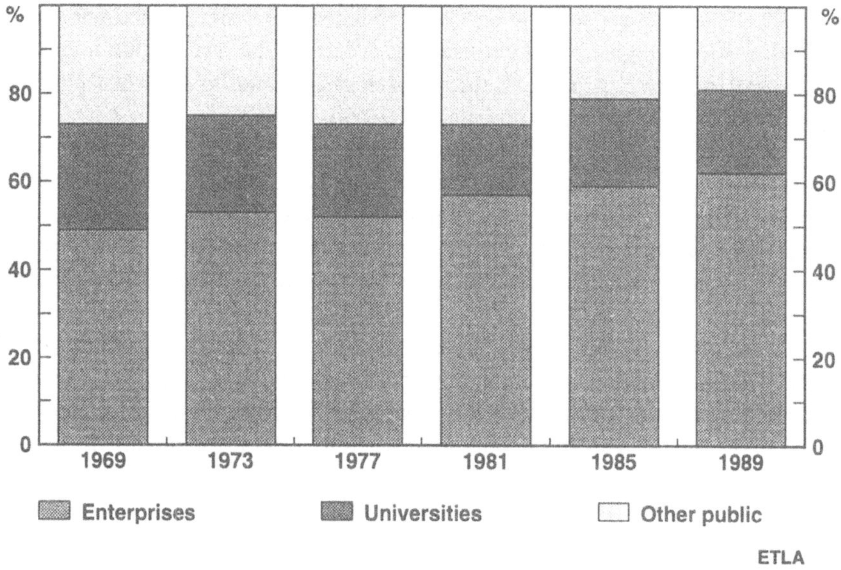

Fig. 8.2. Breakdown of research and development expenditure by performing economic sector in Finland, 1969 - 1989. Source: Statistics Finland.

As can be seen in Figure 8.2, the share of the corporate sector in R&D expenditures in Finland grew steadily from just less than 50% in 1969 to over 60% in

1989. This growth has not been drastic, but it has been quite consistent. In the most technologically advanced OECD countries the share of the corporate sector is about 70%, but in its own category, Finland has stood up quite well to international comparison (OECD 1991).

The self-sufficiency of Finnish firms is shown in Figure 8.3. The share of state funding in corporate R&D expenditures averaged 5% over the 1980s. In the second half of the 1980s, the share of state funding even contracted, even though official policies and measures strongly supported an increase in public support appropriations. The corporate sector's own funding and public support funding both manifested considerable growth during these years.

On the basis of research statistics it has been concluded in various contexts that the share of state funding in Finland is below the average for the OECD countries. This has led to demands on increasing R&D appropriations intended for firms and for the share of public funding in R&D expenditures to be increased to about 10%. These demands have been accepted by the present government, which in other areas has been forced to cut down the support granted to other sectors of the economy.[3]

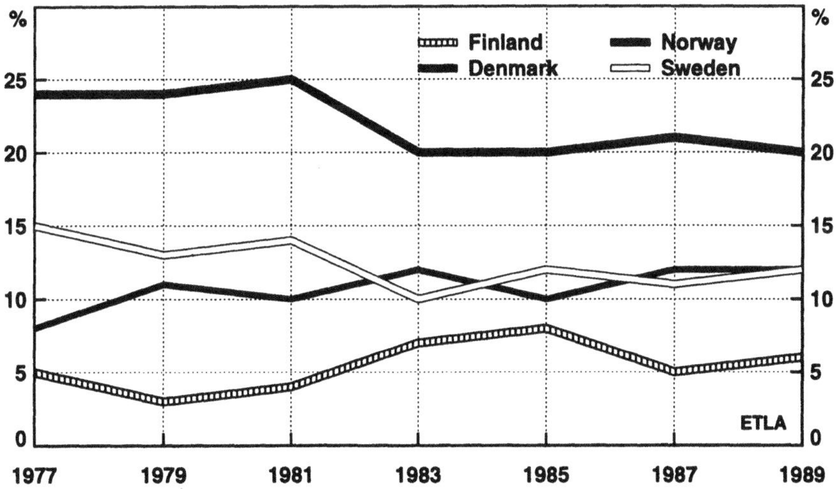

Fig. 8.3. The public funding of corporate research and development expenditure in Denmark, Finland, Norway and Sweden, 1977 - 1989. Sources: Nordforsk and Nordisk Industrifond.

[3] See Box 8.1

Box 8.1 On international comparisons of public funding

Direct international comparisons of the share of public funding are very problematic for the following reasons:

- the statistics include military and civil research
- the terms of public support are different in different countries (loans, venture capital, subsidies)
- there may be significant differences between the institutional classifications of public and private organisations in different countries
- the statistics do not include the public support of R&D via taxation
- statistical errors may be of different magnitudes in different countries (In Finland the figure for public support contained in the research statistics, based on responses given by firms, was just under 50% of the support which, for instance, TEKES and SITRA had granted to firms during the year in question)

Some examples of public funding shares (NUTEK 1992): in 1989 the share of public funding in the R&D expenditures of firms was 30% in the United States, 19% in France, 16% in the United Kingdom, 11% in Germany and 11.5% in Sweden. Of this, the share of military R&D was 100% in the United States, 75% in France and the United Kingdom, 50% in Germany and 65% in Sweden.

Institutional differences and classification procedures explain most of the differences between Finland and Norway. SINTEF, the VTT's counterpart in Norway, and certain other research institutes are classified under the corporate sector in Norway, so the public funding received by these establishments significantly increases the public funding share of the corporate sector. The situation in Denmark is also very similar.

The most important public financiers in Finland in 1989 were (Tilastokeskus 1991a) TEKES 73%, KERA 9%, SITRA 6% and the rest of the public sector 12%.

The corporate sector's degree of self-sufficiency is also indicated by the fact that in 1989 the share of contract research in R&D expenditures was only 5% in the corporate sector, as a whole, and 3% in industry (Tilastokeskus Statistics Finland 1991). These figures included, in principle, all the contract research commissions awarded to the VTT, universities and other enterprises and research institutes in Finland and abroad. The printing industry accounted for the largest share of contract research (21%), compared, for instance, with only 1% for the engineering and electronics industries.

The intensity of state intervention dressed up in the guise of technology policy can also be assessed from the perspective of the share of corporate R&D support in the total amount of public contributions to industries. In 1991 industries (incl. agriculture and forestry) received a total of FIM 13 billion in the form of financial support from the state. Of this amount, primary production and the food industries accounted for 70%, most of which was allocated to agriculture, the manufacturing industries accounting for the rest, 30%.

In addition, it is worth mentioning that it is not industrial R&D but regional policies that predominate the Finnish programmes related to manufacture, followed by export-related support. The share of R&D in the support, amounting to FIM 3.6 billion, was only 10%, i.e. 3% of all support granted to industries (Government report on the management and state of the treasury 1991, Palokangas 1992).[4]

Charles Edquist has concluded (Edquist 1990), that in Sweden the degree of state intervention in the process of technological change is quite limited - and not increasing. This applies to Finland as well.

8.5 The Impact of Technology Policy on the Structural Development of Finnish Industry

The two most notable and, at the same time, most favourable examples of the modernisation occurring in Finnish industry in the 1980s were increased exports of high technology products and the growth of the electrical engineering and electronics industries. The share of high technology products in industrial exports rose from 4% in the early 1980s to 11% in 1990, and the electrical engineering and electronics industries grew by 150% in the 1980s and their share of the metal and engineering industry rose to just under 30% (ETLA 1991, Vartia & Ylä-Anttila 1992).

These two examples of change are closely interlinked. A significant proportion of high technology products belongs to the electrical engineering and electronics industries. The high technology product groups manifesting the strongest growth have been telecommunications equipment and electrical equipment, as can be seen in Figure 8.4.

Figure 8.5 shows that the electrical engineering and electronics industries have had, since the late 1970s a special position in public R&D finance. In 1978 the share of the electrical engineering and electronics industries was 36% and in 1980 as much as 45% of the public finance to industrial projects in the form of risk loans and grants. By the end of 1980s, its share was slightly diminishing, but

[4] According to data collected by the OECD, in 1989 the share of R&D support in the public support received by industry (net costs incurred by the government) in Finland was 11%, i.e. almost the same as the average for the OECD countries (11.5%). The share of R&D support was 56% in the Netherlands, 46% in Denmark, 39% in Japan, 17% in France and 14% in Sweden. The United Kingdom, Austria, Ireland and Norway were on the same level as Finland.

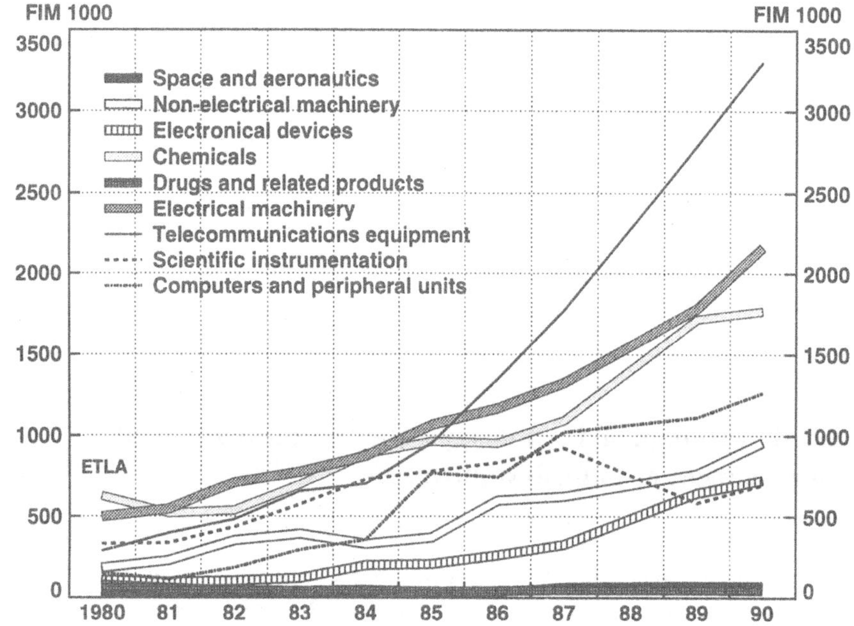

Fig. 8.4. Exports of high technology products by product group, 1980-1990. Source: Statistics Finland.

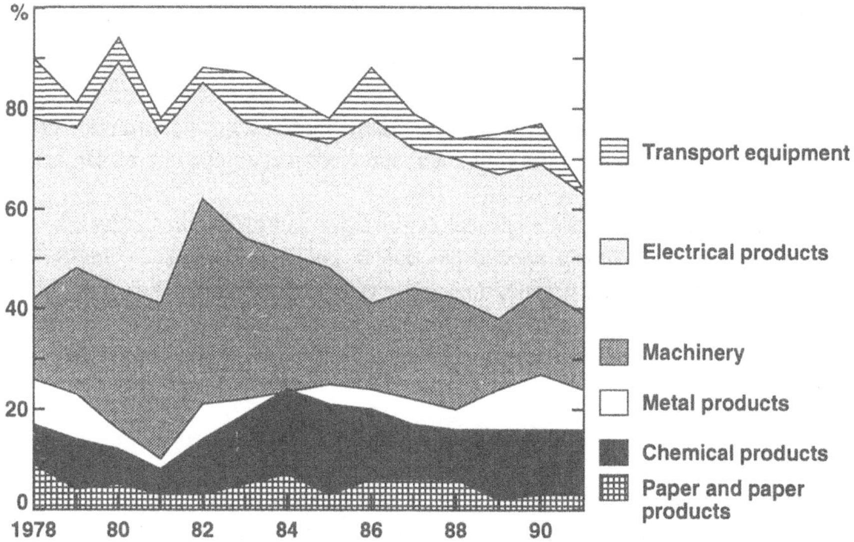

Fig. 8.5. The distribution of funding for industrial research granted by the Ministry of Trade and Industry (1978-1983) and TEKES (1983-) by branches of industry. Sources: Ministry of Trade and Industry and TEKES.

during 1978-1991 around one-third of the funds, on average, went to this branch of industry.

On the other hand, Table 8.2 shows that the distribution of funds by public financiers is very similar to the distribution of the corporate sector's R&D input. If, however, the share of public funding in total R&D expenditures of these economic sectors is examined, it can be stated that, for instance, in 1989 the share of public funding in the research expenditures of the electrical engineering and electronics industries was only 6.3%.

Table 8.2. The distribution of public funding for industrial research and the distribution of R&D within enterprises by the main branches of industry, %.

	1979		1985		1989	
	Ministry	Enterprises	TEKES	Enterprises	TEKES	Enterprises
Paper and paper products	4	9	3	6	2	5
Chemical products	10	16	18	18	14	18
Metal products	9	2	4	2	8	1
Machinery	25	17	23	21	14	12
Electrical products	28	28	25	19	29	26
Transport equipment	5	3	5	7	8	4
Other	19	25	21	26	25	34
Total	100	100	100	100	100	100

Sources: The Ministry of Trade and Industry, TEKES, and Statistics Finland
Ministry = Ministry of Trade and Industry

On these grounds we can assume that the demand of firms rather than anything else have focused the attention of public financiers on the electrical engineering and electronics industries and the engineering industry. A more positive interpretation of the facts could be that the officials responsible for appropriating public funds have been able to adapt themselves to the changing needs of the economy and thus to support more innovative firms and industrial branches.

Information technology was the indisputable focal point of funds for goal- oriented research granted by the Ministry of Trade and Industry and more recently by TEKES in the early 1980s. However, this trend started to change in the mid-1980s. In 1982 the share of information technology was 62%, in 1985 51%, in 1988 32% and in 1991 29%. The VTT also supported the strengthening of information technology through its research programmes and new laboratories, but the shift of information technology from basic technologies to areas of application is also clearly observable within the sphere of the VTT.[5]

The decline in the share was partly due to the fact that the application of information technology shifted to manufacturing technology, process technology and construction. However, there was also an actual decline in the share of information technology, but this was also due to the fact that information technology became the object of criticism in Finland in the 1980s.[6] It was believed in Finland that excessive resources had been directed to fields of advanced technology, which in practice meant information technology or the electrical engineering and electronics industries. It was thought that the emphasis should be shifted towards safeguarding the competitiveness of basic industries and to the application of technology in traditional fields of know-how, e.g. the forest industries and the metal and engineering industries.

The arguments were understandable then, and they are still easy to understand today. In the light of the above, however, one could ask whether the situation with regard to high technology products and the electrical engineering and electronics industries might not be better today if the investments in those areas had been made with even greater intensity and consistency.

8.6 Technology Policy as the Creator of a Favourable Atmosphere

Committees have traditionally played a very important role in the formulation of Finnish technology policy. These committees have formed a cooperation and inter-

---

[5] In the 1940s, electrical engineering research had incorporated with radio engineering research, which, in turn, incorporated with semiconductor research in the 1960s. The VTT's Electronics Laboratory began operating in Oulu in 1974 and the Medical Engineering Laboratory started up in Tampere. The Electronics Laboratory spawned the Computer Technology Laboratory in 1983 and the Optoelectronics Laboratory in 1991. The Laboratory for Information Processing began operating in 1985.

[6] The Committee on the Development of the Technology Programmes has been the principal exponent of the back-to-basics philosophy in Finland. The committee emphasized that the technology programmes should generate product and production technology for the country's basic industries, i.e. the forest industries and the metal and engineering industries. This was, by no means, the first time that advanced technology and the basic industries were seen as both augmenting and competing with one another. The committee wrote down on paper what had been floating around in the air for a number of years previously.

action channel between the public and private sector organisations participating in the formulation and implementation of technology policy. This has been a way of making the goals mutually consistent and committing the various parties involved to attaining the goals. It has also acted as a means of preventing conflicts and contradictions. In other words, technology policy is not just an institution for the allocation of R&D appropriations, but also, to an increasing extent, a social institution.

The Technology Committee (1980) provides an almost classical example of the role played by committees in Finland. This committee convened in the early 1980s and was a particularly broadly based body. Its members included representatives of the government, research institutes, universities, industry and labour organisations. The committee also had links to political decision-makers.

When the committee began its work, its main points of departure were a fear of mass unemployment caused by the information technology revolution, and, consequently, the desire and need to regulate the direction and pace of development by government measures. When, just less than two years after starting work, the committee submitted its findings, its most important conclusion and, at the same time, recommendation was that even rapidly advancing automation would not become a dominant factor in social development during the 1980s, but a resource offering new opportunities. The report gave rise to two different schools of opinion, but neither of these questioned the correctness of this basic conclusion.

The committee not only legitimized the pursuit of technology policy but also gave new or additional justifications for activating technology policy in both the quantitative and qualitative sense (Lemola & Vuorinen 1988). From this well cultivated ground sprouted TEKES and other similar organisations, and the initiation of technology programmes received fresh impetus. Perhaps most importantly, however, the participation of different interest groups in policy formulation gave rise to a broad consensus on the goals for the new decade.

Similarly, the experiences acquired from the technology programmes also indicate that cooperative arrangements, such as these, have direct and many indirect effects on research and technology (Numminen-Guevara 1992). These should be borne in mind when the benefits of the programmes are evaluated. One of the most important indirect effects is the added value which arises when research scientists and users are brought together in a fruitful dialogue and interaction.

8.7 Summary

Increasing the resources made available for R&D or briefly technology push has played a prominent role in Finnish technology policy. The demands for increased resources have been based not only on a belief in the importance of intangible investments as such but also on an awareness that the Finnish companies and the public sector have invested much less in R&D than Finland's main competitors.

On these grounds the support and intensification of R&D in the companies has been seen as a natural and necessary function of the public sector, too. Consequently public appropriations intended to support corporate R&D work and funds for strategic industrial research grew in Finland during the 1970s and 1980s faster than other research expenditures of the state and faster than in the OECD countries, on average.

However, it can be concluded that in Finland the degree of state intervention in the process of technological change has been quite marginal - and is not increasing. This can be seen in the following figures:

- The share of the corporate sector in R&D expenditures in Finland grew steadily from just less than 50% in 1969 to over 60% in 1989.
- The share of state funding in corporate R&D expenditures averaged not more than 5% over the 1980s.
- The share of these state funds for corporate R&D is only 10 % of the public financial support received by industry and services, and if primary production and food industries are included, it is not more than 3%.
- The role of external services or contract research organised by public research institutes and universities has been fairly modest - at least quantitatively - compared with R&D performed by companies themselves.

The electrical engineering and electronics industries or information technologies, in general, have played since the late 1970s a special role in Finnish technology policy. But not even this can be interpreted as a sign of selective state intervention. The distribution of funds by public financiers is very similar to the distribution of the corporate sector's R&D input. The same applies to a very large extent to other public measures as well.

The basic strategy of technology policy, which can be characterised as one of positive or active adaptation was created before the present period of depression. At least until now, the slump has not given any cause to change the thrust of the basic strategy in any other direction. The chief aim in Finland has been to concentrate on safeguarding the continuity of the present instruments and the appropriations intended for their use. Additional appropriations have actually only been allocated to support the R&D work of firms. This has been done without any consideration on potential or desirable new directions to build up economic strength in Finland.

References

Edquist, C. (1990), Technology Policy - Social, Economic and Political Aspects, TEMA-T Working Paper, Linköping.

Freeman, C. (1982), The Economics of Industrial Innovation. London: Pinter Publishers.

ETLA (The Research Institute of the Finnish Economy) (1991), Suhdanne 1991/4, Helsinki (In Finnish).

Falk, P., R. Jaakkola and M. Viikari (1975), "Keskustelu ja ohjelmat", in: Bruun, K (ed.), Tiedepolitiikka ja tutkijan vastuu. Helsinki: Tammi (In Finnish).

Hallituksen kertomus valtiovarain hoidosta ja tilasta 1990 [Government report on the management and state of the treasury] (1991), Helsinki (In Finnish).

Kosonen, P. (1981), Yhteiskunnallisen työnjaon muutos ja pääoman uusintaminen. Helsinki: Helsingin yliopiston sosiologian laitoksen tutkimuksia No. 219 (In Finnish).

Laurila, E. (1982), Muistinvaraisia tarinoita. Keuruu: Otava (In Finnish).

Lemola, T. and R. Lovio (1984), Näkökulmia teollisuuden innovaatiotoimintaan ja teknologiapolitiikkaan Suomessa 1980-luvulla, Valtioneuvoston kanslian julkaisuja 1984:2 (In Finnish).

Lemola, T. and P. Vuorinen (1988), Finländsk teknikpolitik förnyas, in: Glimell, H. (ed.), Industriförnyelse i Norden. Århus: Forlaget Samfundsøkonomi og Planlægning (In Swedish).

Lundvall, B-Å (ed.) (1992), National Systems of Innovation. London: Pinter Publishers.

Luukkonen-Gronow, T. (1975), "Suomen tiedepolitiikan kansainväliset esikuvat", in: Bruun, K. et al. (ed.), Tiedepolitiikka ja tutkijan vastuu. Helsinki: Tammi (In Finnish).

Luukkonen-Gronow, T. (1977), Katsaus muutamien OECD-maiden 60- ja 70- lukujen tiedepolitiikkaan. Helsinki: Suomen Akatemian julkaisuja 12/1977 (In Finnish).

Löppönen, P. and A. Noro (1975), "Tieteen yhteiskunnallisista yhteyksistä", in: Bruun, K. et al. (ed.), Tiedepolitiikka ja tutkijan vastuu. Helsinki: Tammi (In Finnish).

Numminen-Guevara, S. (1993), Katsaus teknologiaohjelmien arviointiin. Espoo: VTT:n teknologian tutkimuksen ryhmän työpapereita 2 (In Finnish).

NUTEK (1992), Forskningssamverkan för förnyelse av näringsliv och energisystem, Stockholm (In Swedish).

OECD (1980), Technical Change and Economic Policy. Paris: OECD.

OECD (1991), Main Science and Technology Indicators 1-2. Paris: OECD.

Ormala, E. (ed.) (1987), Evaluation of Technical Research and Development. Espoo: Nordforsk.

Palokangas, J. (1992), Yritysten valtiontuet Suomessa - tuen kehitys EFTA- ja EY-maissa. Helsinki: VATT (Government Institute for Economic Research), keskustelualoitteita 17 (In Finnish).

Rothwell, R. and W. Zegveld (1981), Industrial Innovation and Public Policy. London: Pinter Publishers.

Sigurdson, J. (1989), "The Changing Scene for Industrial R&D in Europe", in: Fusfeld, H. (ed.), Background Papers for Conference on Changing Global Patterns of Industrial Research and Development, Stockholm.

Teknologiakomitea [Technology Committee] (1980), Teknologiakomitean mietintö, Helsinki (In Finnish).

Teknologiaohjelmien kehittämistoimikunta [The Committee on the Development of Technology Programmes] (1990), Teknologiaohjelmatoiminnan linjat 1990-luvulle, Helsinki (In Finnish).

Tilastokeskus [Central Statistical Office of Finland] (1991a), Yritysten tutkimus- ja kehittämistoiminta 1989, Koulutus ja tutkimus 1990:24 (In Finnish).

Tilastokeskus [Central Statistical Office of Finland] (1991b), Tutkimus- ja kehittämistoiminta 1989, Koulutus ja tutkimus 1991:3 (In Finnish).

Valtiontalouden tarkastusvirasto [The State Audit Office] (1991), Teollisuuden tutkimus- ja tuotekehitystoiminnan edistäminen (In Finnish).

Vartia, P. and P. Ylä-Anttila (1992), Kansantalous 2017. Helsinki: ETLA (The Research Institute of the Finnish Economy), Series B No. 81 (In Finnish).

YTK-komitea (YTK Committee) (1974), Yritysten tutkimus- ja kehitystoiminnan edistäminen, Helsinki 1974 (In Finnish).

9 Comment on Tarmo Lemola's Article

David Roessner
School of Public Policy, Georgia Institute of Technology, USA

Max Jakobson, in the beginning of his little book *Finland: Myth and Reality*, notes that Finland pays a price for her policy of neutrality: the rest of the world has little incentive to follow Finnish affairs; their knowledge of the country tends to be superficial and fragmentary. "As a result, Finland is forever at the mercy of the itinerant columnist who after lunch and cocktails in Helsinki is ready to pronounce himself upon the fate of the Finnish people" (p. 8).

So it is with a great deal of humility that I offer comments on a paper whose subject is technology policy in Finland, not only because I know so little about the country, but also because technology policy and innovation are themselves such enormously complex subjects. Indeed, years of studying technology policy and innovation in my own country leaves me with a deep appreciation for the difficulty of diagnosing problems associated with technological change, much less prescribing policy solutions for them.

The study on which this book is based acknowledges the complexity of technology policy by appropriately characterizing the complex of activities that technology policies seek to influence as the "innovation system." Christopher Freeman defines the national innovation system as "the network of institutions in the public and private sectors whose activities and interactions initiate, import, modify, and diffuse new technologies" (Technology Policy and Economic Performance, p. 1). The scope and complexity of this range of activities speak to the difficulties the science and technology policy analyst faces.

Despite all these obstacles, I think there are potential contributions an outsider can make to analysis of national innovation systems. I recently had an opportunity to experience this first-hand when I presented a seminar on U.S. technology policy at the Fraunhofer Institute for Systems Technology and Innovation in Karlsruhe. Following my portrayal of the many weaknesses in the U.S. innovation system, and the inadequacies of our country's recent technology policies, the German analysts reminded me of some of the strengths of our system (e.g., mobility of professionals across institutional lines) that most other countries do not enjoy. The perspective I gained from these "outsiders" proved helpful, and so I hope I can do the same here. My approach will be, first, to share with you some features of the Finnish system that were surprising to me, an outsider, and which I thought would have significant effects on science and technology policy. Second, I will narrow the focus of my remarks to Mr. Lemola's article, but retain the perspective of the surprised outsider, reacting as much to what is *excluded* from his article as to what is *included.*

My sources for the first part of my remarks are the introductory chapter to this book; ETLA's *The Finnish Economy, 1992/3;* the 1987 ETLA Discussion Paper by Pentti Vartia and Synnöve Vuori, "Development and Technological Transformation - The Country Study for Finland"; Severi Keinälä's "Finnish High-tech Industries and European Integration," and Pentti Vartia's "New Technologies and Structural Changes in a Small Country".

First, the number of key economic players, especially in technology-intensive industries, is very small. I noted, for example, that the fifteen companies that spend more than FIM 100 M constitute about 80% of all private expenditures on R&D (Keinälä 1990, pp. 96-97), and that the ten biggest manufacturing companies (four of which are state-owned) account for more than half the R&D, and 44% of manufacturing exports and total manufacturing sales (Vuori and Vuorinen 1993). The main products of five of these ten are forest products and (apparently) related machinery (Ibid., and Vartia and Vuori 1987, p.34). Nokia alone accounts for 22% of total industrial R&D.

In itself, this is not surprising in a nation of five million people. What I find surprising about this is that its implications for technology policy do not seem to be reflected in the analyses I read. For the most part, these analyses are macroeconomic in level and overall approach. Data are aggregated by industrial sector or product code, yet the technology strategies and economic fortunes of virtually a handful of companies will dramatically alter the results of a particular public program. The decisions of, say, a dozen people would appear to me to determine whether a program of R&D subsidies or a tax credit succeeds or fails, yet analysis seems to stop at the boundary defined by industry or product code - and never approaches the walls of the individual firm.

Second, the analyses I read that seek to diagnose the problems of the Finnish manufacturing sector, and to describe the associated public technology policies involved, focus almost exclusively on the "input" end of the innovation process. It is as if the analysts and policymakers had accepted the pipeline model of innovation: if enough R&D is done, targeted at the right industrial sectors, productivity growth will result.

This, too, might not be so surprising, especially to an American presumably imbued with the idea that only market failure can justify government action in technology policy, and that market failure can be identified readily only in the conduct of long-range research. But in the case of Finland you have no equivalent fears of government action, and you would appear to be a country likely to be characterized by Henry Ergas as "diffusion-oriented" rather than "mission-oriented." In the former, national sovereignty is the goal, and R&D decision making is focused on a few industries dominated by large firms, while in the latter the emphasis is on investment in human capital and strengthening industrial associations; technology policy is "an intrinsic part of the provision of innovation-related public goods" (Ergas, "Does Technology Policy Matter?").

In Finland I expected technology policies to build on the idea of "created comparative advantages" you already possess: strong physical infrastructure, high levels of education, very homogeneous population, ease of communication among institutions and between suppliers and final manufacturers. In several places brief mention was made of what I expected to be the major foci of both technology policy analysis (diagnosis, prescription) and technology policy itself (programs):

- "... the main strength of Finnish high-tech industries must be, and in most cases is, specialized excellency, implemented in high-quality products that follow close behind the most advanced technological development, complemented by niche marketing" (Keinälä 1990, p. 54)
- diffusion of flexible manufacturing systems (FMS) is particularly rapid in Finland because of strong social infrastructure and high levels of education, "a prerequisite for adoption and efficient running of FM systems" (Vartia 1990, p. 8).

So, I was left with a bit of a mystery: with some evidence that Finnish high-tech companies seem to follow a diffusion-oriented, market-niche strategy, why do Finnish technology policy and policy analysis apparently fail to complement this emphasis?

Now to Mr. Lemola's article. His purpose is to determine the effect that Finnish technology policy has had; his criterion is the effect that public R&D has had on the innovative activities of firms. The policy to be evaluated began in the 1970s and consists of three "pillars" (p. 186):

- direct support of corporate R&D via product development loans and tax relief;
- reorienting activities of public R&D;
- support for "goal-orented technical research".

As I mentioned earlier, this is clearly a "technology push" policy although, as Mr. Lemola points out, this strategy is more pragmatic and small-scale than comprehensive and strongly interventionist.

As you might expect, given my previous remarks, I found it interesting to contrast Mr. Lemola's definition of technology policy as policy whose key objectives are to "create new corporate activities and support innovation in existing firms" (p. 184) with his statement that the effectiveness of technology policy "largely depends on how well R&D subjects are selected and how well measures are planned and implemented." The definition encompasses the entire range of innovation from research through manufacturing and marketing, while his criterion of effectiveness for such policy seems to focus only on the "inputs" or "upstream" aspects of the innovation process. While in the case of Finland this narrower focus is at first glance appropriate, it leads to an evaluation of what *was* done to influence innovation rather than what was *not*.

Mr. Lemola's summary of evaluations of public R&D programs in Finland yields a commonly-observed conclusion: measures of effectiveness that suggest a mixed bag of effectiveness, with a healthy dose of uncertainty because of the complexity of innovation processes and the stringent (yet necessary) criterion of a program's value: the *independent* effect of public action. Despite the necessity of drawing cautious, heavily circumscribed conclusions, Mr. Lemola nevertheless makes two important points:

- There exists a strong positive association between recent patterns of public R&D investments and those of industry (thus destroying the critics who claim that public R&D has been misdirected);
- The Technology Committee is a significant institution in Finnish technology policy. It is the kind of social institution which to me has the potential to take advantage of Finland's small size and homogeneity to formulate technology policy informally and on a consensus basis.

Mr. Lemola questions whether "positive adaptation" will be enough for the 1990s, or whether efforts should be made to promote restructuring more purposefully and at the same time more selectively. I would urge Mr. Lemola to be even bolder and suggest that technology policy for the 1990s focus on the entire innovation process, especially manufacturing and marketing, and dare to base some of its programs on knowledge about how innovation is managed within the walls of the firm. This is particularly relevant in view of the relatively new "culture of R&D and innovation" in Finnish companies, as mentioned in this morning's discussion. Doing so will, I think, suggest some public initiatives whose targets are well beyond R&D.

In closing, I would like to suggest modestly that new pillars be added to the "three pillars of Finnish technology policy" mentioned in Mr. Lemola's article:

a fourth pillar: support public and private R&D intended to strengthen the ability of companies to follow a "second to market" industrial strategy;

a fifth pillar: strengthen the diffusion-oriented infrastructure elements that support flexible, high quality manufacturing by both large and small firms, their suppliers;

a sixth pillar: support analysis, information outreach, and information access activities that would assist companies in a selective, market-niche product development strategy.

Thank you very much for allowing me the opportunity to share these thoughts with you.

References

The Finnish Economy (1992). Helsinki: ETLA (The Research Institute of the Finnish Economy), 1992/3.

Ergas, H. (1987), "Does technology policy matter?", in: Dasgupta, P. and P. Stoneman (eds.), Economic policy and technological performance. Cambridge: Cambridge University Press.

Freeman, C. (1987), Technology policy and economic performance. London: Frances Pinter Publishers.

Jakobson, M. (1987), Myth and reality. Keuruu: Otava.

Keinälä, S. (1990), Finnish high-tech industries and European integration. Helsinki: ETLA (The Research Institute of the Finnish Economy), Series C No. 60.

Vartia, P. (1990), New technologies and structural changes in a small country. ETLA (The Research Institute of the Finnish Economy), Discussion Papers No. 317.

Vartia, P. and S. Vuori (1987), Development and technological transformation - The country study for Finland. ETLA (The Research Institute of the Finnish Economy), Discussion Papers No. 245.

S. Vuori and P.Vuorinen (1993), "Outlines of the Finnish Innovation System: The Institutional Setup and Performance", first chapter of this volume.

10 The Rigidities and Potential of a National Innovation System

Synnöve Vuori and Pentti Vuorinen
ETLA, The Research Institute of the Finnish Economy, Helsinki, Finland

10.1 What Was This All About?

This book deals with technological change. We have discussed it from the perspective of a small (and, in the European perspective, a rather peripheral) country, Finland. The Finnish conditions clearly differ from those of larger and more central countries with longer industrial traditions. On the other hand, many basic aspects of technological progress are similar everywhere and independent of country size.

In the economic literature it is typical to present and compare national economies with statistical macro-level indicators. However, highly aggregated macroeconomic figures on industrial and technological progress only give a vague and superficial picture of the differences. Their power to describe - and even less to explain - the dynamics behind the developments is weak.

In this book, we have especially tried to tackle some aspects of the dynamics of technological advances in the Finnish industry. Further on, we also try to put our arguments in a comparative perspective. We have used the conceptual framework of the 'national systems of innovation' (NSI) as a starting point. However, we do not give much space for theoretical discussion, and all the chapters are based on the individual thinking of the authors.

In the first chapter we describe the Finnish innovation system, the institutional set-up and performance. This chapter also served as a common information base for the authors of the other articles. The task of the other authors of the book was then to produce their own explanation for the developments that took place. The chapter contains a description of where we are now: what did the technological and economic development in Finland look like in the last few decades? What kind of framework produced these developments, that is, what are the essential characteristics of the Finnish innovation and production system? In the last couple of decades research efforts have grown faster than in most other OECD countries, and many favourable results have been reached: increased productivity, more high-tech exports and more international patents. Except for a few narrow areas, however, technology gaps still exist in comparison to many advanced countries.

The results of the intensive use of technological inputs in the 1980s still remain to be seen. Technology diffusion - using innovations created elsewhere and adapting them to one's own needs - is an extremely important dimension of technological change in a small country (see e.g. Vuori and Ylä-Anttila 1992). In fact, a remarkable share of research efforts may just entail keeping up with developments in other firms, research institutes and countries.

In the early 1990s, large changes in economic prospects, industrial structures, the welfare society, as well as international cooperation and trade have been eroding the formerly solid ground of the national innovation system. In this new situation the basis for this system must evidently be redesigned. With increasing internationalization, the possibilities for a truly national innovation system become more limited. For Finland, the role of the forest block remains central in the structure of the economy, and thus it faces a great challenge in staying technologically and economically competitive.

10.2 Dimensions and Problems of the Finnish Development

In his article *Allardt* (Chapter 2) puts technological progress in the context of the many dimensions of modernization. One of his themes is the interplay between social and technological changes. Can rapid technological renewal and change within the economy take place without corresponding social changes? As one example of lagging social change he discusses the rigidities created by the powerful interest organizations. The rapidly grown legal web of entitlements and rights have become an obstacle for many social and economic advances; they also seem to be one of the reasons behind the present overall economic problems. Allardt also takes the goals of technology policy under discussion: when almost one fifth of the labour force is unemployed, should not employment creation be the selfevident number one goal of technology policy?

According to *Kasvio* (Chapter 3), Finland is a somewhat of a special case among the Nordic countries. Her industrial modernization started later than in other West European countries, and after the Second World War she has had a rather specific relationship to the former Soviet Union. Also the institutions of an advanced welfare state and corporatist industrial relations developed more slowly than in the other Nordic countries. Finland allocates a larger share of its public spending to education nowadays than any other OECD country, but this does not guarantee the system's superior performance. At present, the entire position of Finland in the post-war European system is changing. Radical changes are needed also within the social infrastructures of production. The Nordic countries will probably move from the present forms of 'democratic corporatism' into the predominance of 'enterprise-level corporatism' or 'microcorporatism', in which wage negotiations are decentralized and the most essential decisions are produced within the enterprises.

To adjust to a world where an increasing number of corporations no longer have a distinguishable nationality, the role of national policies has to be reconsidered. An overall institutional restructuring of the Nordic societies into well-functioning post-Fordist network economies requires creating social infrastructures that actively support the adoption of innovative practices in all spheres of life. One opportunity for positive developments could be to participate in the development of the region surrounding the Baltic Sea. If the problems threatening this area can be solved, one could in the future perhaps talk not about national innovation systems, but of a dynamically developing regional system of innovations ranging from St. Petersburg to Northern Germany.

Kanniainen (Chapter 4) addresses the role of collective risk sharing and analyses what modern growth theory can tell about the reasons for growth. He concludes that the high growth rates of the Finnish economy in the last few decades can at least partly be explained by the growth incentives created by the public sector and by the success of the forest block, and to some extent also by the catching-up phenomenon. While the growth process has been strongly policy-related in the positive sense, economic institutions and especially the extensive public insurance of economic risks has diminished the need to face risks and to manage them properly. At the same time as public risk-sharing has promoted R&D efforts, this may have been offset by public measures which have adversely influenced the structure of investment. Kanniainen sees a need to evaluate more deeply the trade-off between social returns and social risks of various policy measures and also of their incentive effects on the private sector.

Pessi (Chapter 5) looks at the working of the national innovation system from the point of view of corporate management. Factors which have positively affected developments in the Finnish industry include a favourable company culture, flexibility as well as strong technological performance and manufacturing-orientation of companies. On the other hand, their market orientation has not been strong enough, which is partly a consequence of the long period with intensive clearing-based countertrade with the former Soviet Union. Growing globalization and increasing economic integration will inevitably change this situation. Another unfavourable factor has been the development of the capital structure of companies, including a relatively low share of equity capital, and limited possiblities for obtaining venture capital. An area calling for more emphasis is enhancing the possibilities for technology transfer, for instance by promoting the start-up and performance of cooperation networks.

Gregersen, Johnson and Kristensen (Chapter 6) compare the national systems of innovation in Finland, Denmark and Sweden. They deal with aspects of the institutional set-up and the economic structure which strongly affect the processes of interactive learning, and identify the most important industrial development blocks. While sharing many common characteristics, these three small countries are by no means similar. All of them have strong resource-based development blocks, but the performance of the innovation and production systems differ. Sweden clearly outperforms the other two countries technologically, but its economic performance is not as impressive. On the other hand, Denmark with relatively low technological performance, a large share of small firms and strong emphasis on the agro-industrial sector continues to perform economically rather well. Finland seems to fall between. Therefore, the conclusion is that the relationships between 'immediate performance' and 'ultimate performance' are not very straightforward.

Carlsson (Chapter 7) partly tackles the same issues when comparing Finland, Sweden, Japan and the United States, but uses the concept of technological systems (TS) as the basis of his analysis rather than national systems of innovation. In his view, Sweden's relatively poor growth record is due to problems in the domestic environment which are more related to institutions than technology: the welfare state with high taxes and poor incentives, too central a role of non-market principles, corporativistic structures, and lack of EC membership. These thoughts are interestingly parallel with those of Kanniainen (cf. above) concerning Finland.

According to Carlsson, Finland has a robust technological system in its forest-based industries and parts of the engineering industries, but is becoming increasingly dependent on collaboration with foreign firms. Finland appears to need an oversize technology base to constitute an attractive base for both domestic and foreign firms. On the other hand, new business formation has developed much more favourably than in Sweden.

Japan's strong performance until recently seems to be based on well-functioning technological systems in the area of mechatronics. Carlsson notes, however, that in other industrial sectors and several service industries the record is much less impressive. In the United States the technological systems in the engineering sector have been strong, largely as a result of the 'military-industrial complex'. Despite weaker economic performance lately, results in the areas of innovation and higher education have continued to be strong. The U.S. also seems to be strong in non-engineering technologies, such as software and biotechnology, and in many rapidly growing service sectors.

What has been the role of technology policy for the performance of the Finnish innovation system? According to *Lemola* (Chapter 8), increasing resources for R&D has been the main goal of the policy. Despite rapid growth of public appropriations for corporate R&D efforts, the degree of state intervention in the process of technological change has been limited, and is not seen to be increasing. The electrical engineering and electronics industries have since the late 1970s received much emphasis, but the public financiers just seem to have reinforced the distribution of funds which would have developed in the corporate sector anyway. The basic strategy of positive or active adaptation has not changed during the economic crisis of the early 1990s, and additional funds have only been directed to support the R&D of firms. There has been no strong urge to find new directions for technology policy.

Roessner, in commenting on Lemola's article (Chapter 9), suggests three additional 'pillars of technology policy'. The first one involves supporting public and private R&D intended to strengthen the ability of companies to follow a 'second to market' industrial strategy. The second consists of diffusion-oriented infrastructure elements, and the third pillar involves supporting analysis, information outreach, and information access activities. Roessner also notes the very dominant role of a few large companies in creating technological change in Finland.

10.3 The NSI Concept

There is not much theoretical discussion in the book. The focus is more on making some experiments with the already relatively well-established concepts on the Finnish reality. However, a few issues can be picked up from the articles.

First: there is a need to clarify the relations to parallel approaches. Porter's cluster analysis comes quite near to the National Systems of Innovation (NSI) framework, and so do various theories of business systems as well. In his article, Bo Carlsson takes up the concept of technological systems (TS) as a competitor of the NSI approach. However, his argumentation shows that the differences are mostly a question of a different focus. TS has technology and technology-based relations as the core issues, and national economic relations and development are seen from the angle of technology diffusion, infiltration and adaption. The NSI approach, on the other hand, puts the focus on national economic dependencies and examines technological change in this context. Thus the two approaches are in fact quite complementary. When combined, they could offer an interesting, matrix-resembling framework for studying economic and technological change.

Second: the relations of various components within the NSI need to be clarified. Learning through user-producer relations and other forms of close cooperation throughout the national system has traditionally been taken as the core of the system. It has quite often also been the main subject in empirical, micro-level studies based on the NSI framework. In connection to this, formal systems for training and education are always mentioned as being among the most important institutions within the NSI.

The actual role of the formal educational system, the organizational forms and educational practices of the system and the interaction between learning by doing and formally acquired qualifications have nevertheless gained much less interest. For example, Kasvio (cf. above) notes that Finland allocates the largest share of its public spending to education within the OECD. This does not guarantee, however, the system's superior performance. There are problems in the match between the educational system, the needs of companies and the actual learning practices within the companies.

This seems to be true in respect of the practical role of many other social, political and cultural institutions as well. The rigidities of the social institutions created during the 1970s and 1980s have become obstacles for economic and technological development. The changing role of labour market organizations, both on the national and the shop floor level, mentioned by Allardt and Kasvio, are another example of this. However, little empirical - and even less internationally comparative work - exists for instance on the effects of labour market organizations on innovativeness and the national technological development paths. This means that much more work is needed to connect the institutional framework more closely to the micro-level perspective, firm behaviour and innovative practices.

The particular need for studying more deeply the interaction of the overall institutional framework and micro-level practices is emphasized by the present, both internationally and often nationally unstable and rapidly changing economic and social situations. In the Finnish case, special demands are put on the ability of the institutional framework to change in accordance with the needs of completely new economic and international set-ups as compared with the mid-1980s. It is not only a question of the flexibility of the national institutions to follow external changes, but of the ability of the institutions to grasp future needs and to be able to create corresponding conditions.

In this respect, Kanniainen takes up an interesting issue in his article. The tradition of high collective sharing of risks has had a long lasting effect on the Finnish system. This seems to be true with respect to security and risk sharing systems for both firms and persons. Of course, this is not uncommon in the other Nordic welfare states either.

The issue comes up in the present unfortunate economic situation and shows many weaknesses. The comprehensive collective risk sharing system cannot remain in an economic environment of slower growth, higher unemployment and higher needs for flexibility and adaptability. The question is, however, to what extent - and in what areas - is it favourable to dismantle this system of collective security. Where would more risk be favourable for economic, technological and social development, and where would it only bring along more uncertainty and change the focus toward more short-term developments?

There is an international dimension in this as well. Would some kind of more international risk sharing and security systems be able to compensate the national systems, brought by for instance further European integration? And what issues would need this kind of safety net? How strong is, actually, the difference between ensuring social and economic security to people and building up a collective risk sharing system for business risks?

In these remarkably changed conditions, the authorities in charge of technology policy are also facing the need to redirect their measures. Much more than before, technology policy must be seen as one part of the more encompassing social and economic policies highly dependent on the conditions provided by the other sectors of policy for the functioning of individuals, firms and the whole society. The role of factors such as well-functioning institutions, appropriate education and sound incentive systems cannot be too much emphasized. Resources devoted to R&D and the diffusion of technologies must be seen as complementary elements of the support to technological change. In addition, support measures should be fairly selective and also provide opportunities for opening new areas of industrial development, even though this occasionally will lead to failures. The broader perspectives are in fact starting to be seen in public measures. In the recently published national industrial strategy for Finland (see National ..., 1993), industrial policy is seen as consisting of several broad areas, and developing the national innovation system is taken as the starting point of technology policy.

10.4 Further Issues to Study, National and Conceptual

Finland has been able to raise her technological level considerably in the past few decades, but a gap with respect to the most advanced countries still exist. Moreover, the base of technological progress is still narrow. It should also be remembered that a few large firms account for the majority of R&D, for example. Also in Sweden large firms dominate, but in Denmark the role of small firms is much larger.

The micro-level processes of innovation and learning within business firms are not discussed much in this book. The actual user-producer relations, company networks and interactive learning processes have, as of yet, not been studied sufficiently in Finland to allow international comparisons, although some interesting empirical studies are going on. Evidently more research should be done on these issues, both in the context on the national innovation system and in an internationally more comparative research set-up.

One of the main findings of the Finnish national system of innovation and production was the thinness of development blocks. The forest industrial block is by far the most important, the actual core of the whole national system. This block has been relatively well studied, but the linkages outside, the opening up of the block - as well as other blocks - and creation of further 'spin-offs' to other economic areas are still relatively unknown areas. One, quite ad hoc example is the interplay between the forest industry and textiles: With growing environmental consciousness and higher oil prices artificial fibres become less advantageous. Wood-based fibres such as viscose might offer a good alternative. However, most viscose raw materials are imported to Finland. The opportunities to develop this dimension of the 'forest industry' have been neglected, although textiles and clothing industries have been relatively much more important than in for instance Germany, which supplies most of the viscose materials to Finland.

This is only one example of the possible close-ups of the development blocks. The national focus has, perhaps, been blinded by the success of the core trajectories of the development block during the two previous decades and thus closed the view to alternatives. This seems to be true with respect to many environment related issues within the Finnish forest industrial block also: the industry became conscious of the issues only relatively late, and is still somewhat lagging in comparison to many competitors, though the industry has gained remarkable success especially in making the processes less pollutant.

The point is that even the 'environmentalization' of the industry has occurred in line with the long-term technological, economic and institutional development path of the Finnish forest industry. More radical innovations and new directions of technological progress may have been hampered by the strength of the established institutions - and ways of thinking - within the block. In this respect, more focus should be put on (seemingly) more peripheral and new issues also within the old development blocks.

The difficulties of comparing even culturally close national systems are stressed recurrently in Lundvall's (1992) book on national systems of innovation. The comparative articles in our book can only confirm this notion. While it is relatively easy to compare economic systems and economic performances, it is already problematic to put formal social and cultural institutions on the same comparative line. Things become even more complicated, when we have to study the differences in firm behaviour, learning processes and actual cultural practices in real organizations. However, more concrete comparative studies are needed, and we think that the articles of Gregersen et al. and Carlsson give a good starting point from which to continue comparing national systems of innovation.

Small countries share many common characteristics, but still there are large differences between them. This is true even within the culturally and historically close Nordic countries. Sweden is in general very different in terms of performance, even though the infrastructure and general set-up of the societies and economies are fairly similar in the Nordic countries. However, Finland and Sweden seem to share the same kind of encompassing collective risk-sharing systems, which may have detrimental effects on incentives related to innovativeness also.

Comparing the outcomes of national innovation systems in the article by Gregersen et al. shows a new interesting aspect, that is well worth developing further. They take up the actual main question of all economic development often forgotten in the context of technology and innovations: what is the goal of economic and technological development and how well have different national systems succeeded in achieving them? It is, however, evident that various countries also define the final goals of their techno-economic development very differently.

10.5 National, Regional or International Systems of Innovation?

Lundvall (1992) strongly emphasizes the role of national and cultural institutions in techno-economic development. The national and cultural aspects have often been pointed out as an explanation for the success of small, ethnically homogenous countries like Finland. However, the small countries have also been more dependent on the international economy and international relations. This dependence is growing rapidly because of technical and global economic and institutional developments - such as European integration. The internationalization of the Finnish innovation system has also been very rapid, after the slow start.

Will the role of the national framework become weaker? Or is it just the other way round: have the more closed national systems reached their peak, and only to decline without stronger foreign influences and connections?

Lundvall (1992) warns about the growing uncertainties and difficulties in communication between partners with very different cultural backgrounds. He sees the danger of slowing down of technological development because of more cautious behaviour caused by cultural uncertainties.

However, the process of internationalization and integration is also as such a process of learning, a process of institutional innovation. It is quite likely that the roles of national - in many cases even regional or local - innovation systems and closer international cooperation will find different forms. On the one hand, the requirements for international cooperation are evident especially in radical innovations and innovations needing a very wide market. On the other hand, many single, concrete innovations created through daily user-producer linkages still need a very close cooperation that is much easier to create between companies with similar cultural background and a common national framework. This is true in the case of most everyday technology diffusion to (and between) small and medium-sized firms also.

The emergence of new regional constellations for economic and technological cooperation is also very likely. Examples of regions - not in the subnational, but in the international sense - with dynamic growth exist: the Pacific Rim, the Alpine region and so on. Kasvio mentions growing cooperation within the Baltic Sea region as having the potential to make the north-eastern periphery of Europe into a new growth centre. The existing technology gaps and intense drive to develop functioning market economies in the area support the developments, even though the immense problems of the collapsed Eastern European system hamper the birth of subtle economic institutions.

In addition to the changes required by the internationalization process, the current economic crisis in Finland provides another challenge: how can the Finnish national innovation system adjust to the new situation with very little room for growth, and even a need to cut expenses and lay down activities which were earlier deemed necessary? It is important to see beyond the present difficult times so that the fundamentals of the innovation system are not destroyed, after which the reconstruction process would be extremely costly. A process of creative destruction is needed, especially to break the rigidities created during the past couple of decades, but the sound basis and dynamic core of the system must not be broken. Some kind of dynamic innovation system is needed, be it regional, national or international, and even more so to create opportunities for a better and more prosperous future.

References

Lundvall, B.-Å. (1992), "Introduction", in: Lundvall, B.-Å. (ed.), National systems of innovation - Toward a theory of innovation and interactive learning. London: pinter Publishers.

Vuori, S. and P. Ylä-Anttila (eds.) (1992), Mastering technology diffusion - The Finnish experience. Helsinki: ETLA (The Research Institute of the Finnish Economy), Series B No. 82.

National industrial strategy for Finland (1993), Helsinki: Ministry of Trade and Industry, Publications 1993/1.

GPSR Compliance

The European Union's (EU) General Product Safety Regulation (GPSR) is a set of rules that requires consumer products to be safe and our obligations to ensure this.

If you have any concerns about our products, you can contact us on ProductSafety@springernature.com

In case Publisher is established outside the EU, the EU authorized representative is:

Springer Nature Customer Service Center GmbH
Europaplatz 3
69115 Heidelberg, Germany

Batch number: 10725973

Printed by Printforce, the Netherlands